William Russel Dudley

The Cayuga Flora

Part I: A Catalogue of the Phænogamia Growing without Cultivation in the Cayuga

Lake Basin

William Russel Dudley

The Cayuga Flora
 Part I: A Catalogue of the Phænogamia Growing without Cultivation in the Cayuga Lake Basin

ISBN/EAN: 9783337269951

Printed in Europe, USA, Canada, Australia, Japan

Cover: Foto ©berggeist007 / pixelio.de

More available books at **www.hansebooks.com**

MAP OF
THE LAKE REGION
OF
CENTRAL NEW YORK.

SCALE OF MILES.

LAKE ONTARIO

Mathews, Northrup & Co., Art Printing Works, Buffalo, N. Y.

ULYSSES
LANSING
ENFIELD
DRYDEN
ITHACA
CAYUGA
FOREST HOME
VARNA
Renwick
S. H. Glenwood Pt.
Enfield P. O.
Pleasant Grove Cem.
Turkey Hill
Woodlawn Cem.
North Quarry
Judd's Falls
McKinney's Twin Glens
Dryden-Lansing Swamp
Cliff Park
Cascadilla
Indian Cr.
Williams Br.
Coy Glen
Lake Ridge
S. H.
Cem.
Ch.

BULLETIN OF THE CORNELL UNIVERSITY.

(SCIENCE.)

VOL. II.

THE CAYUGA FLORA.

PART I : A CATALOGUE OF THE PHÆNOGAMIA GROWING
WITHOUT CULTIVATION IN THE CAYUGA
LAKE BASIN.

BY

WILLIAM R. DUDLEY.

ITHACA, N. Y. :

ANDRUS & CHURCH,

1886.

The title, "Bulletin of Cornell University, (Science)," is here revived with the permission of the authorities of this University, and out of regard and affection for the memory of Professor Charles Frederick Hartt, whom the earlier Cornell students remember as a man of brilliant and captivating intellectual qualities, and a source of inspiration to thorough and untiring work in Science ; who projected a series under the above title, and with his friend and pupil Professor Derby, published the initial volume in 1874, containing papers on the geology and palæontology of Brazil.

Preface.

The local explorations of this region, which have resulted in the accumulation of the facts recorded in the present publication, were begun more for the pleasure they involved, than with any well-defined purpose in view. Soon, however, they were taken up more earnestly, with the desire of enlarging our knowledge of the immediate resources of the Botanical Department of Cornell University. Since 1880, there have been added objects, viz. : to ascertain as precisely as possible the abundance of each species, the local distribution of it, and the number and habits of those forms whose limit of general distribution on this continent falls within or near our territory, all, of course, involving the taking of voluminous field-notes, and the collection of a large series of specimens. But it must be remembered that while the results of observation on geographical distribution are recorded, the catalogue is primarily intended to assist the student and local collector, and contains many things which would be of little interest to the non-resident. For instance, the dates of the local flowering of plants are given with considerable closeness ; these are for average years, and due allowance must be made with the spring flowers, for such a season as the present (1886), when everything since the first week in April has been two or three weeks in advance of the average season ; and naturally these dates might not answer for a climate differing from ours. Again, localities are often mentioned, for plants not especially rare, solely for the purpose of aiding student collectors. Forms and varieties of species have been included in the catalogue numbers, with somewhat more than ordinary fullness, in order to call forth further observations from local and other collectors, on their abundance and constancy. We have been chary of giving new names to such, in order to avoid a possible increase of synonyms, although we believe some of those numbered will prove to be, when better known, well-marked varieties.

Acknowledgements are due from the writer to several of his botanical friends for pronouncing upon varying or critical forms. Dr. Gray, in the midst of most arduous labors, kindly went over a large package of specimens sent him, contributing valuable notes and suggestions, especially on *Ranunculaceæ* and *Compositæ*. Moreover he most thoughtfully and opportunely forwarded for use, the proof-sheets of his "*Revision of the North American Ranunculi.*" Mr. Sereno Watson, of Cambridge, assisted me on the *Rosaceæ* and *Apetalæ*. Our *Carya sulcata* was submitted with many other woody plants to Professor C. S. Sargent. Dr. Vasey of the National Herbarium has seen a large number of our grasses and given me much information on forms collected here and elsewhere. He examined a large number of Eatonias in the herbariums of the eastern cities before naming the herein-described species of that genus which had long interested me. To the herbarium of Mr. M. S. Bebb, of Rockford, Ill., has been contributed at different times, full sets of our interesting *Salices*. In return he has contributed very valuable notes, drawings and specimens, and our correspondence has been a most delightful one. In spite of ill-health he has freely advised me ; and defining the position of the hybrid willows is chiefly his work, any opinion of my own having in most cases, received his sanction. Mr. Morong, of Ashland, Mass., has equally favored me in his especial field the *Naiadaceæ*. His contribution of specimens to my own herbarium has been of very great value. Valuable information or specimens have also been received from Dr. Charles H. Peck of the N. Y. State Herbarium, from Mr. E. L. Hankenson, of Wayne Co., and from Professors Trelease, J. M. Coulter, and L. H. Bailey.

In making explorations and amassing the materials and facts for the catalogue, the writer has of course devoted much of his own time and resources, and has for certain periods privately employed the assistance of special students ; but he here wishes to make a general acknowledgement for information always freely reported,—when anything interesting turned up,—by the special students in botany and others, whose helpful and generous ways have often been the chief encouragement in the pursuit of the work.

The work of exploration, collection, revision, critical study, correspondence and the accumulation of collateral facts, has all been carried on as a secondary matter to unusually arduous duties of instruction. It is believed, however, that the work has been thorough enough to justify publication, and the reader is assured that all localities given in the catalogue are selected from records written where the plants grew, or made at the end of the excursion on which they were observed. A full, frank statement is made where necessary, of observations, of opinions held in regard to distribution, and of other matters, in order to furnish an unequivocal basis for emendations, corrections and additions,—as well as to give information,—believing that a mere list of names, not easily capable of verification, really furthers our knowledge but little. It is to be hoped that this contribution will be met in the spirit in which it is offered, its oversights and manifest shortcomings pardoned in the endeavor to contribute additional, authentic information, whenever such is obtained by those using the catalogue. Any one finding plants new to the region or new stations for rare plants, should send us information of the locality, accompanied if possible by the specimens, and all such information will be put on file at the Botanical Laboratory, in the future, as it has been for several years past. It is proposed that whenever such a list accumulates sufficiently to warrant it, to publish a supplementary slip containing the additions, with the names of the discoverers, and to send it to any one who may desire it.

During the collection of materials for the phænogamic list, a large number of the higher cryptogams have been pressed, and the material largely worked up; also a considerable number of parasitic fungi have been saved and much work already done on them, and at no distant day we may hope to place before our students something tangible on several of the more important groups of lower plants.

The author has recognized more clearly than any one else could, that this work is pioneer work, but resolved that it should be done with a certain degree of thoroughness, believing that it is only by full and accurate records of this kind, made for every section of the country, that we shall ever succeed in demonstrating beyond dispute the great and profoundly interesting problems of geographical distribution of species. Nevertheless, he feels that this list is merely a contribution to the better knowledge of our plants; and it does not for a moment pretend to approach the position of a completed history of their distribution in this region. Although we have explored most of the apparently interesting nooks within the limits, still there are probably others unobserved, as there are also known localities deserving further attention. And when we recognize the fact that after centuries of close scrutiny some corner of Great Britain periodically discloses a plant never before seen on the island, and that since the publication of the Flora of Washington in 1881, scores of new discoveries have been made within its limits, we know that many new ones await us here.

Botanical Laboratory,
Cornell University, May 20, 1886.

Introduction.

Limits of the Flora and its Physical Characters.

This catalogue professes to include all the flowering plants, so far as observed, growing without cultivation in the territory drained by Cayuga Lake, its tributary brooks and creeks. It also includes the plants of the water-shed marshes and ponds lying between our hydrographic system and the neighboring ones,—besides the altogether independent, outlying and exceedingly interesting little water-system of the West Junius ponds. The latter, an important addition to our illustrative resources, belongs as much to our system as to any of the smaller ones, and is therefore combined with it in this work. The water-shed marshes and ponds sometimes drain into the Cayuga system as in the case of Locke Pond, Dryden Lake, the Round Marshes, the White Church and Brookton Springs; sometimes away from it, as in the case of Cayuta Lake, Michigan Hollow Swamp and the Fir-Tree Swamp in Danby, and the larger Marl Ponds in South Cortland. There is an outflow in freshet seasons only, from Summit Marsh, and from Freeville bog and the Fir-Tree Swamp at Freeville, and then into other streams than ours, while the pond and marsh near Chicago Station, also several of the Marl Ponds have no apparent outlet.

The greatest length of this territory is about sixty-five miles, extending southward from Montezuma, on the Erie canal, to Summit Marsh, in the northern part of Tioga County. Its average breadth could be considered about eighteen miles. Its least breadth, near the northern marshes, is only from four to six miles. From this region the basin gradually widens till near the vicinity of Ithaca it suddenly expands, becoming on the line of its greatest breadth about thirty-two miles. This line extends from Cayuta Lake on the southwest, through Ithaca to the Marl Ponds on the northeast. In this part of the Cayuga basin occur the larger streams, such as Fall Creek, Cascadilla, Six Mile and Neguaena Creeks, all entering the "Inlet" and the south end of the lake in a group. The only rival streams are Salmon Creek and Taughannock Creek, entering the lake about ten miles from the southern end. All of these streams flow through preglacial valleys, the larger of which their slender currents have altered but little from the ancient form. Of this part of the basin, Ithaca is the central point, and its county (Tompkins), is wholly within our limits. Considerable portions of Cayuga Co., on the eastern shore, and of Seneca Co., on the western shore of Cayuga Lake, and small fractions of Schuyler, Tioga and Cortland, also fall within this basin.

The central topographical feature is Cayuga Lake; and, indeed, by reference to the map it will be seen to be the central feature of the whole lake-system of Central New York. The neighboring lakes are all at a higher level, and pour their waters, either into the vast level marshes which are manifestly but a northern continuation of the great Cayuga valley, or into Cayuga Lake itself. Seneca, Owasco and the other lakes usually possess a well-defined northern shore, but the Cayuga Marshes, raised but little above the level of our lake, so blend into its shallows that its exact northern termination is difficult to define; and they give to it,—at least to that portion of it—a character entirely its own, which was recognized by the "Six Nations," or ancient Kanonsionni, who called what we now know as Cayuga Lake, "Tiohero,"[1] the *lake of flags or rushes*, or *lake of the marsh*.[2] The limit of the Flora is therefore extended down these marshes to Montezuma. The length of the lake is usually estimated at 38 miles, its breadth from 1½ to 3 miles. In appearance therefore it resembles a great river; indeed it is said to occupy a part of a preglacial river channel of which the Neguæna valley was the continuation. The height of the lake above mean tide is 383 feet,[3] the greatest depth found by numerous soundings of the Cornell University Engineering Department was 435 feet, at a point directly off Kidder's Ferry. In the section between Myers Point and Sheldrake Point it is in many places over 400 feet deep. On account of its depths its waters are comparatively cold far into the summer, and rarely become so chilled in winter as to admit of the formation of ice over the deeper sections. From one-half to two-thirds of the middle section usually remains open, but in the winter of 1884–5 the lake was frozen over before the middle of February and the ice did not break up till the first week in April. There is a tradition that this occurs about once in twenty years. The temperature of the lake unquestionably influences the development of vegetation in its immediate vicinity. Plants on its shores are usually a week later in the Spring than in the neighboring ravines and the warm valley about Ithaca, and a week earlier than on the distant highest hills; and during the first half of November, the blue flowers of *Aster lævis* and the white plumes of *Aster sagithfolius* still remain in considerable abundance, while they have long ago matured and faded near Ithaca. The extremes of the natural climate are in this manner so modified that the eastern slopes of Cayuga and the other lakes have always grown the

---

[1] See *Relations des Jésuites*, for the year 1672, Quebec Ed., III, p. 22; also the map in the same volume.

[2] See Notes of Gen'l J. S. Clark, in the Journal of Lieut. Hardenbergh, p. 71.. The name is variously written *Thiohero*, *Tichero*, *Choharo*. At the time of the establishment of the Jesuit mission in 1656, at the foot of the lake, and for a century after, the name was also applied to one of the three principal towns of the Cayuga tribe of the great League, viz.: that on the eastern side of the Cayuga Marshes near where the turnpike crosses the outlet at present. Apparently the same word with same meaning—written *Deyohero*—is in the Canienga or Mohawk dialect. See Hale's explanation in his *Iroquois Book of Rites*, 1883, p. 121.

[3] See Survey of the Geneva, Ithaca and Sayre R. R.

peach[1] successfully. The lake waters are above the average in purity
excepting at the extreme northern and southern extremities, as was
demonstrated a few years since by analyses made by my friend Professor Breneman.

Proceeding southward from the gently sloping shores near Cayuga
Bridge the banks become gradually bolder, until in the vicinity of
Levanna the first cliffs appear on the eastern shore. Between Willets
and Kings Ferry these reach their culmination in the "High Cliffs;"
but stretches of lofty, precipitous, or more or less broken declivities
occur on both shores until within a few miles of the southern extrem-
ity. At intervals especially near the mouth of some stream are low
half-sandy points which yield many rare plants. Near Ithaca, and
about two miles from the lake, the great valley forks, the main portion continuing to the right of South Hill, a preglacial valley of ero-
sion extending southwardly to Waverly in the Susquehanna valley.
The other portion on the left of South Hill is similar to the first and
forms the present Six Mile Creek and White Church valleys, and
opens into the Susquehanna at Owego. These deep valleys penetrate
and cut through the great dividing ridge between the St. Lawrence
or Great Lake hydrographic system to which our streams and smaller
lakes are tributary, and the Susquehanna system, and are parallel to
similar valleys east and west of us. The head waters of the streams
occupying them, i. e., the summits between the two systems are
usually very near the crossing of the dividing ridge. The points of
greatest elevation in our whole region, the most precipitous, as well
well as botanically the most interesting of our inland declivities, and
the watershed marshes, springs and ponds being in fairly close con-
tiguity, as the following statements will show.

The subjoined stations with elevations above mean tide, recorded in
the current reports of the N. Y. State Survey by the numbers preceding each, are upon this dividing ridge southwest of us, and include the
highest points yet measured in that direction:

No. 405, Urbana sta., Steuben Co., west of Hammondsport, 1940 feet.
    " 401, Hornby sta.,   "   "   " "   "   2045 "
   " 394, Sproul sta., Schuyler Co., west of Watkins,   2091 "
   " 226, Orange sta.,   "   " southwest of Watkins,   2033 "
   " 428, Couch sta.,   "   " southwest of Cayuta L.   1679 "
   " 403, Newfield sta., Tompkins Co., Saxon Hill.   2095 "

Cayuta Lake, one of our watershed lakes, lies only about two miles
southwest of Saxon Hill. The latter point can be readily discerned .
from the University Campus, as the highest elevation of the farther
rim of blue hills, and lies a few points south of west. This rim of hills
is the dividing ridge above spoken of. It trends eastward, crossing

---

[1] In the Sullivan campaign against the Six Nations in 1779, Col.
Butler destroyed an Indian town on the present site of Aurora called
Chonodote or Peach-tree Town, together with "about 1500 peach trees
besides apple trees and other fruit trees." Even as late as 1845,
Schoolcraft found "within the boundaries of Aurora, the remains of
an apple orchard, which was ruthlessly cut down by a detachment of
Sullivan's army," *Notes on the Iroquois*, p. 57. The apple trees and
probably the peach trees were introduced by the Jesuits.

the Neguæna valley, south of west Danby, where a spur of it on the eastern side, forms the interesting knob of Thacher's Pinnacle, and a mile farther east appears in a still more elevated point, Ball Hill, its height unknown but certainly rivalling the Caroline hills. At the eastern base of Ball Hill lies Michigan Hollow Swamp, about 1400 feet above tide, while two or three miles south of where the ridge crossed the Neguæna, is Summit Marsh, near the source of Neguæna Creek. The ridge continues eastward to Durfee Hill, west of the White Church valley and reappears on the east side in the bold cliffs of the North and South Pinnacles and then rises to the rounded summits of Bald Hill[1] and Taft Hill, usually known as the Caroline hills. Northeast between the Caroline hills and the Virgil hills lies Dryden Lake.

The following records from the Survey of N. Y. show some of the elevations along the divide east and northeast of us, the first named being on a northerly somewhat isolated branch :

No. 113, Dryden sta., on Creamer Hill, central part of Dryden, 1880 ft.
"   385, Richford sta., Tioga Co., east of Caroline Hills,   1995 "
"   402, Virgil sta., Cortland Co., east of Woodwardia Sw.,   2133 "
"   419, Morehead sta.,   "   "   east of the Round Marshes, 1865 "
"   382, Solon sta.,   "   "   northeast of Cortland,   1977 "
"   107, Niles sta.. Cayuga Co., east of Owasco L.   1621 "

It will be useful to compare the elevations of some of the marshes, ponds, "summits" or R. R. stations in the valleys near the dividing ridge. These are derived chiefly from the R. R. Surveys.

W. Danby railroad station	872 feet
Summit " "	1072 "
Summit Marsh	1050 "
White Church railroad station	958 "
Dryden Lake, (surface.)	1160 "
Summit of the S. C. R. R.	1215 "
Freeville railroad station	1049 "
Malloryville "	1057 "
McLean "	1090 "
Chicago "	1169 "
South Cortland, (Marl Ponds)	1151 "
Cortland	1116 "
The Round Marshes	1100± "

The four higher sphagnum swamps in our basin are the little Spruce Swamp on the Enfield Hills, less than two miles south of the N. Y. S. S. 399, and approaching it in elevation, Michigan Hollow and Fir-Tree Swamps in Danby, and the swamps about Locke Pond. The latter is probably the most elevated pond in our region. The lowest sphagnum swamps are Larch Meadow and Fleming Meadow about 400° above the sea.

The following are the elevations of certain subordinate, but interesting points :

[1] The height of Ball Hill, according to Professor H. S. Williams is about 1900 feet. Taft Hill has about the same elevation.

Ithaca	392 feet
No. 399, Enfield sta., (N. Y. S. S.)	1464 "
" 420, Turner sta., " South Hill, Ithaca.	1125 "
" 425, Hungerford sta., " Eagle Hill, " . .	1273 "
Cornell Univ. S. S., (the central sta. of our map)	788 "

The station number is always plainly marked on the small granite monument erected at each of the stations of the N. Y. Survey.

From this rapid review of the territory comprised in our Flora it will be seen that it has a peculiar form, and varies much in height, ranging from 383 feet at the lake-level, to 2095 feet on the southwestern and 2133 feet on the eastern boundaries. Nevertheless these heights can scarcely be called mountains,[1] but giant hills instead, scored out of an ancient table-land. They rarely present cliffs or exposed rocks, excepting where some recent ravine or a preglacial valley has been cut through the undisturbed strata, but are rounded and cultivated or at least tillable to their summits.

There remains but one other feature to mention in this general review. Nothing in the physical aspect of this region strikes the stranger as more characteristic than the so-called gorges or ravines, found in the first great bench above the lake and valleys, wherever a creek or even brook descends to the lower level. The true gorges are probably without exception, of recent or post-glacial origin, the walls are frequently of perpendicular or overhanging rock from fifty to two hundred feet, or even much higher, as in Taughannock and Enfield ravines. Within these great chasms are usually falls or cascades, some of them exceedingly beautiful and of considerable height.

The ravines themselves are often flanked on either side by a succession of well-marked terraces of sand or gravel, the most conspicuous cases being near Coy Glen, Enfield, and Buttermilk ravines, clearly visible many miles away.

For comparison with other regions the following elevations are given, derived from railroad, canal and official U. S. Surveys :

Canandaigua L.	668 feet	Skaneateles L.	932	feet
Keuka L.	718 "	The Tully Lakes	1200	"
Seneca L.	445 "	Lake Erie	573	"
Cayuga L.	383 "	Lake Ontario	247	"
Owasco L.	713 "			

The summit level of the canal south from Seneca L. is 884 feet.

The topographical features of this region have so modified or controlled the distribution and occurrence of plant forms that the brief outline of them given above seemed absolutely necessary. In other words, soil and exposure have been chiefly influential in this distribution, but in certain species, the connection of the plant-habitat with the great and more remote centers of distribution, together with the elevation, high or low as the case may be, furnish important, perhaps essential, conditions. It will be seen that the southern part of our

---

[1] On the map accompanying Kalm's Travel, published 1772, the heights southeast of Cayuga L. (probably the Caroline hills) are styled the "Onugarechny Mts."

Flora occupies a portion of that singular and much eroded table-land which forms the extreme outworks of the great Alleghany mountain-chain, but which is cut off from it by the rather narrow valley of the Susquehanna River. This fact presents us with the interesting and subtle problem of our relationship to the peculiar Alleghanian flora, for the solution of which our section of the table-land ought to furnish a better field than any other, both on account of its elevation and the narrowness at this point of the valley separating us from the Alleghany foothills. Furthermore this table-land was the boundary of the ancient inland sea, which according to a favorite hypothesis with geologists, once existed over the whole area occupied by the great lakes and the low country adjoining, and which may have drained southwardly through several of the deep channels, the so-called preglacial valleys, continuing southwardly from most of our Central New York lakes. If this hypothesis is correct the North should have contributed some peculiar plant to our region as it has to Niagara, the present gateway of its waters. It is reasonable to suppose that a few would become permanent if the conditions were favorable. To the first of these problems we are not fully prepared to give an answer although we have visited the Alleghanies in Virginia and elsewhere for comparison of floras. Upon the second, more can be said and both are touched upon later in the discussion of the " Lesser Local Floras."

The Maps.

The Map of the Lake Region of Central New York.—The title of this map sufficiently explains its chief object. It was also designed to show the relation of our own hydrographic basin to those surrounding it, and to furnish a convenient hand-map to those in adjoining counties, who are making field observations in botany. There is little about its construction that professes to be original. Within our own basin, however, the interesting marshes and higher hills and certain features of Cayuga Lake have been indicated, in addition to those given on the ordinary maps. The maps of N. Y. are notoriously inaccurate[1] as regards the exact geographical position of towns, and must be much more so in regard to streams, features always slurred over by commercial map-makers. As the above-named map was based on the French map and those of Asher and Adams' Gazetteer, the errors, except certain prominent ones in our own basin, could not be eliminated.

The Map of Ithaca and Vicinity.—This is designed for the use of collectors who wish to go out for a single day's excursion from the University. In its construction we took as a basis the excellent map of Cayuga Lake and its shores, constructed after a careful survey by the Engineering Department of this University, and we would thank the professors in that department not only for the use of this, but for additional measurements derived from the same source. Of the vari-

---

[1] See statement in Report of the N. Y. State Survey for 1876 and 1877, by JamesT. Gardiner.

ous local maps, we found that the maps showing the most nearly correct survey were those in the Atlas of Tompkins Co., by Stone and
Stewart. These agreed so well with Professor Fuertes' map in respect
to the ground covered by both, that the Atlas was used as a guide in
completing our map. The Atlas is almost the only map giving the
north line of Ithaca township correctly. The position of roads, railroads, schoolhouses, marshes, etc. were verified by personal observation, and town-boundaries by inspection of original records. The
task of correctly representing the streams was considerable, as the existing maps were full of errors regarding them. All we can now say
is, that after visiting these and making careful observation, we corrected a large number of errors and omissions, and our map is much more
nearly accurate than the older ones. What every one must look forward
to with expectancy, is the completion of the admirable State Survey
now in progress under Mr. James T. Gardiner. With that as a basis
there will be a possibility of making wholly accurate local surveys.
It is believed, however, that such results will bring about few relative
changes in our local map. We therefore feel justified in introducing
in it a mechanical arrangement by means of which, as applied to this
map, the exact location of any natural object may be briefly indicated
on any label or permanent record in connection with our botanical or
other natural history surveys. After various less satisfactory schemes
to meet a long felt need of the writer, and his colleague, Professor
Comstock, the latter devised a plan much more simple and useful
than the ordinary system of squares applied to maps, and at the writer's request he drew up the following explanation of it, which was
published in *Science*, Vol. VII, (1886), p. 352. "For the purposes of our
local survey a well-known point on the University grounds is taken
as a centre. Upon a map of this locality, a north and south line and
an east and west line are drawn through this point. These lines are
marked o. Other lines are drawn parallel to these lines dividing the
map into squares, each line indicating the distance of one kilometre.
These lines are numbered, beginning in each case at the one next the
zero line and reading toward the margin of the page. By means of roads,
streams and other conspicuous objects, the position on the map, of any
locality can be easily ascertained ; and its distance north or south of
one zero line, and east or west of the other, seen at a glance. It is
only necessary to write figures indicating those co-ordinates upon a
printed blank label to accurately indicate the locality. This label
should have printed upon it the name of the centre of reference ; it
may also have letters indicating two of the cardinal points of the
compass. In the latter case four sets of labels would be necessary.
The following is an example :—

Cornell U. } This filled out might read as follows: { Cornell U.
   N.   E. }   { N. 23, E. 16½

This system was suggested to me by the way in which localities are
indicated in the city of Washington."

The central point taken for our map, and one whose latitude and
longitude are accurately known, is the Cornell University S. S., locat-
ed a short distance south of Morrill Hall. It makes the University

the center of all our future explorations and field observations in this region. Although the English measure has been used on the larger map, in the descriptive portion of this introduction, and in the elevations given, because all previous estimates have been in this measure, the Kilometre[1] is used on the smaller map. The metric system is used in the body of the catalogue and wherever practicable, for all our permanent scientific records.

In the catalogue, instant apprehension is desirable rather than brevity in the record, therefore the above system of abbreviations is but sparingly used to indicate localities. Certain woods and other localities, however, are often referred to but are not named on the map. We give these below:

Bates Woods, N. 4½, W. 4½.   Negundo Wds, S. 2½, W. 3.

Dart Woods, S. 3, E. 7.   "Nook" greenhouse, N. ¾, W. ¾.

Eagle Hill, S. 2⅛, E. 2¾.   Renwick Farm, N. 1 to 3, W. 0 to 5.

Ferris Brook, (mouth) S. 1¼, W. ¾.   Rhodes Woods, N. 1½, E. S..

Fleming S. House, S. 3½, W. 3¾.   Stevens Woods, N. 2, W. 3¾.

Glen Pond, S. ½, W. ½.   Valentine Brook, S. 1½, W. ½

Land-slide (S. M. Cr.) S. 2½, E. ⅝.   Valley Cem., S. 4, W. 4¼.

McGowan Woods, S. ½, E. 2¾'   Willow Pond, S. ⅝, E. 0.

The necessary reductions and the draughting of the maps were principally accomplished by my friend and pupil, Mr. O. E. Pearce of the present Senior class, and the Department of Palæobotany at Washington. To his knowledge of botany was added familiarity with technical drawing and his faithful work I cannot too warmly commend.

The Lesser Floras.

Naturally a larger Flora is made up of distinct regions each possessing its characteristic plants. They might be classified as follows:

1. *The Montezuma and Cayuga Marshes.* For convenience we might say the beginning of these marshes was at Farley's Point and Canoga, for here we first strike the peculiar plants of the marshes. These are *Hibiscus Moschentos*, *Dianthera*, *Cyperus Michauxianus* and *C. Engelmanni*, *Spartina* or Cordgrass, *Panicum virgatum*, and formerly *Hippuris* and *Rhexia*. In addition to these, at Cayuga Bridge, appear *Phragmites* or the "Reed," *Nymphæa tuberosa*, *Mikania*, *Wolffia* and *Carex alata*. For the rest these marshes are made up chiefly of flags, sedges and coarse grasses, the most common being *Typha*, *Scirpus fluviatilis* and *Deyeuxia Canadensis*, which cover the perfectly level, prairie-like expanse for fifteen miles. It is the salt-springs and brackish soil about them, however, which afford the most unique group. The old Indian salt-springs described by Father Raffeix,[3] west of Cayuga Marshes and north of the Demont bridge, give us *Eleocharis rostellata;* and the several abundant springs at Montezuma, one near the canal and two near the banks of Salt Creek, together with their brackish meadows produce the following:

---

[1] The Kilometre = 0.62138 mile.

[2] Elmwood Glen, Camp Warwick and Franklin's ravine, three ravines, either with summer camps or with cottages, occur in the order named north of King's Ferry, Cayuga Lake.

[3] See "Primitive Flora."

Ranunculus Cymbalaria
Chenopodium rubrum, (*Blitum maritimum*)
Juncus Gerardi
Zannichellia palustris
Scirpus maritimus
Panicum proliferum
Diplachne fascicularis, (*Leptochloa*)
Atriplex hastata, (with very red stems.)

We only need to find *Ruppia* in Salt Creek, and *Salicornia* and *Triglochin maritimum* to make our saline list equal to that of Onondaga Lake, as given by Clinton.

2. *Cayuga Lake Shore.* —The most characteristic plants of the lake cliffs extend only into the lower sections of the ravines at or near the head of the lake. A few are found fringing the rocks of the Negaena and White Church valleys. Among the former are *Shepherdia, Symphoricarpus racemosus*, var. *pauciflorus, Neillia, Myosotis verna* and *Salix longifolia*. On the low sandy points on the lake shore, particularly on the limestone is found *Juncus alpinus*, var. *insignis*, while near by may be found *Astragalus Canadensis, A. Cooperi, Vicia Americana, Lathyrus palustris* and its var. *myrtifolius, Aster Tradescenti* and *Polygonum lapathifolium.* A group of introduced shore plants is interesting. It includes *Lythrum Salicaria, Scleranthus annuus, Linaria Elatine, Bidens bipinnata, Scabiosa australis, Lactuca Scariola* and *Polygonum nodosum*, the latter possibly native.

3. *Cayuga L. its Inlets and Outlet :* The aquatics of the lake are unusually luxuriant, and the forms perhaps more than ordinarily numerous. Those not found elsewhere in the Flora are *Myriophyllum verticillatum, M. heterophyllum, Naias marina* and its *varieties gracilis* and *recurvata.* The *Potamogetons* are especially luxuriant and abundant.

3. *The Alluvial Flood-plains and Creek-bottoms.*—These are the most sheltered spots in our whole Flora. The principal one is the Ithaca plain and its characteristic region is Negundo Woods, in aspect a bit of western river-bottom woods. Small ones occur at the mouths of ravines along the lake and even within the ravines. The soil is usually deep, rich and productive though sometimes gravelly. A few very rare plants belong to these levels, among them the more southern species of *Chærophyllum*, Box Elder, the Kentucky Coffee-Tree ; also *Arisæma Dracontium* and *Echinocystis* are found here. The Hackberry, the Downy and the Green Ash, the Bur Oak and Silver Maple, *Alnus serrulata, Staphylea, Mertensia Virginica* are characteristic of the long deep groove of the Cayuga Lake and Negaena valleys but in rare cases they stray out of it. Here also are found all our willows—species and hybrids, excepting *Salix myrtilloides*, although they are not all confined to the alluvial tracts ; indeed no genus of plants so predominates in the valley excepting some of the *Gramineæ.*

5. *Ravines.*—These, a unique feature of our region, often bring together the plants of deep swampy woods, those of the alluvial bottoms, those of the exposed cliffs of the lake, those of the cold wet

cliffs of subarctic latitudes, and lastly the plants of the dry woods, usually found near the brink of the ravine, suggest those of the dry ridges of the Alleghanies; the latter we will place under the next heading. While the ravines shelter the rare *Clematis verticillaris*, *Quercus prinoides* (tree form) *Populus balsamifera*, *Polymnia Uvedalia*, *Blephilia* the two species of *Lophanthus*, *Monarda clinopodia*, *Carex Steudelii*, and a host of our common but most beautiful spring and summer plants, their most remarkable inhabitants are the plants of the wet cliffs few in number, but forming a significant group. They are *Pinguicula vulgaris*, *Primula Mistassinica*, *Saxifraga aizoides*, all growing together at Taughannock, the first two in Fall Creek, the first in Cascadilla ravine. They are always on the south wall, excepting one north side station in Cascadilla for *Pinguicula*. For *Pinguicula* and *Saxifraga*, our stations seem to be more southern than any other in America. *Primula* is found at the head of L. Keuka, whose latitude is a few miles south of that of Ithaca. To be classed with these is *Draba arabisans*, occuring in two small ravines on the east shore of Cayuga L. It is found in northern N. Y. and in Akron, Erie Co., N. Y., considerably north of our latitude. All the above are plants of British America extending to Hudson's Bay and Alaska.

6. *The Woodlands.* —The dry woods of this region, often appearing somewhat barren, really present us with an interesting group of plants. They occupy the knolls along the borders of our ravines, such as Fall Creek and Buttermilk Creek, but their peculiarities appear more striking perhaps on the so-called "pinnacles" of W. Danby and Caroline. Their characteristic trees are the Pitch Pine, Rock Oak Yellow Oak, occasionally the Scarlet Oak and Pignut Hickory. For undergrowth *Vaccinium stamineum*, *V. Pennsylvanicum* and *V. vacillans*, *Rhododendron nudiflorum*, *Ceanothus*, *Solidago squarrosa* and *S. bicolor*, *Gerardia pedicularia*, *G. quercifolia* and *G. flava*, *Gillenia trifoliata*, *Salix humilis*, *Panicum nervosum*, *Asclepias phytolaccoides*, the *Desmodiums* and *Hieracium venosum*. *Kalmia latifolia*, *Epigæa*, *Chimaphila umbellata*, *Chamælirium*, *Pyrola rotundifolia*, also occur, usually in woods somewhat more moist. The rare plants are *Pterospora*, *Oryzopsis Canadensis*, *Panicum xanthophysum*, *Deyeuxia Porteri* and *Lonicera glauca*, var. (the form 402 of the Cat.) *Myrica Comptonia* is occasional.

Of the damp woods, those crowning the higher hills, in some cases pass into those just mentioned, but present in addition a sprinkling of more northern plants. *Habenaria orbiculata* will be found in them sparingly and in woods of this character on the summits of the high hills of Danby and Caroline. occur our only specimens of the rare *Pogonia verticillata*, none of our specimens yet found in flower.

Passing over the woods of beech and maple abundant in the eastern the eastern section of our basin, also the swampy woods and swamps of black-ash and elm which cover extensive tracts north of Forest Home we shall find in the swampy woods of hemlock, often surrounding the sphagnum swamps of Dryden, etc., some beautiful species. In the drier portions are *Viburnum lantanoides*, *Trillium erythrocarpum* *Cypripedium acaule*, *Cornus Canadensis*, *Clintonia*, *Viola rotundifol-*

ia ; while among the green, luxuriant mosses of the low places occur *Cypripedium pubescens*, *Mitella nuda*, *Habenaria fimbriata*, *Coptis*, *Calla*, and rarely *Corallorhiza innata*, *Microstylis monophyllos* and *Trillium cernuum*. There is a gradual transition from swampy woods of this character to the deep swamps containing sphagnum.

7. *Sphagnum Swamps and Open Peat-bogs.*—Of the first there are fifteen or more, including the springy tracts where sphagnum occurs in some abundance. None of these are extensive, and many are overgrown with *Nemopanthes* and *Vaccinium corymbosum*, or in a few cases with Tamarack and Black Spruce. In openings in these we find the royal flowers of *Cypripedium spectabile*, also *Chiogenes* the *Eriophorums*, *Drosera*, *Habenaria dilatata*, *Spiranthes Romanzoffiana*, and often great masses of *Trollius*. Some of the same species occur in the sphagnum about the remarkable springs which gush out in great numbers from the drift-banks or the terminal moraines near Mud Creek, near Malloryville, near the Round Marshes and Rake Creek, in the valley north of Dryden Lake, near Brookton and W. Danby.

The Open Peat-bogs are few in number and are gradually passing over into the sphagnum swamps ; the Freeville Peat-bog, the Malloryville Marsh and several of the Round Marshes, Larch Meadow,— now drained,—the pond-marsh southeast of Chicago sta. belonging to this type. The Round Marshes, two miles east of McLean, are the most interesting of the group mentioned, principally because the fine old woods of hemlock and birch surrounding them have been most happily preserved by the owners. The marshes consist of a chain of five small peat-bogs, separated by swampy tracts, also a half-sphagnum, natural meadow with sedges, orchids, *Menyanthes* etc., surrounding a large spring-pond. Bordering the meadow are a considerable number of the *Populus balsamifera*, *var. candicans*, probably planted by the first settler of McLean whose cabin was built near the spring in the notch hard by. North, through Gracie's swamp, runs Beaver Creek derived from the numerous springs of this region. The plants of the Open Peat-bogs are better marked as a group, than those of any of the other lesser Floras. They are too well known to need mention, and chiefly belong to the *Ericaceæ*, *Orchidaceæ*, *Cyperaceæ*, and are mostly of northern origin. *Sarracenia* occurs in all and *Menyanthes* in nearly all, *Salix myrtilloides* and *Arethusa* are peculiar to the Freeville bog and *Eriophorum alpinum* to the Round Marshes.

8. *The Water-shed Marshes and Ponds.*—These marshes often exhibit a combination of the floral characters of the several marshes and swamps described. So far as our present knowledge goes the following plants are peculiar to their respective localities : *Potamogeton Spirillus* to Cayuta L., *Potamogeton obtusifolius* and *Glyceria Canadensis* to Summit Marsh, *Rhododendron maximum* and *Goodyera Menziesii* to Michigan Hollow, *Carex capillaris* to the Marl Ponds, *Myrica Gale* to Locke Pond, beside those mentioned at the end of the preceding section.

9. *West Junius.*—Newton's two ponds, Lowery's two ponds and a fifth west of these, lie in a sandy rolling country of entirely different

character from the Ithaca region. They are in reality very large springs from which are always pouring large brooks of cold water. Their united waters flow north to the Clyde River. Beyond a sandy ridge and a few rods south of the ponds, is a Tamarack swamp extending to the Pout Pond, the outlet of which is Black Brook flowing south into the Seneca River, west of Waterloo. These are, therefore, watershed marshes and ponds half-way between the Cayuga and the Clyde River water-systems. The following plants are found in the marshes about these several ponds,—excepting the last eight, which occur in damp sandy soil not far away,—and have not been observed in the Cayuga Flora proper :

Drosera intermedia, var. Americana.	Scleria verticillata.
	Carex Buxbaumii.
Valeriana sylvatica.	" chordorhiza.
Solidago Ohioensis.	" decomposita.
" neglecta, var. linoides.	" disticha.
Utricularia cornuta.	Tephrosia Virginiana.
Gentiana linearis.	Baptisia tinctoria.
Bartonia tenella.	Kalmia angustifolia.
Habenaria blephariglottis.	Pycnanthemum lanceolatum.
" ciliaris.	Cyperus filiculmis.
Triglochin palustre.	Carex monile.
Scheuchzeria palustris.	Panicum microcarpum.
Rhynchospora capillacea.	Dicentra eximia?

Gerardia purpurea, Gentiana crinita, Thuja, Habenaria tridentata, Arethusa, Eleocharis rostellata, Cladium, Carex alata, Corydalis glauca, Galium pilosum, Andromeda ligustrina, Myrica cerifera, although elsewhere in our limits, are rare, and are inhabitants of West Junius. The ponds and their marshes where occur this odd combination of Atlantic coast and northern species, are like a little section from Bergen Swamp, Genesee Co. Considering their very limited area they contain as interesting a flora as any other region in our state.

The Affinities of the Cayuga Flora.

Without doubt it is closely allied to the plants of that vast area including the Ohio valley, the Great Lakes and a part of Canada. Our territory of course, skirts the southern or southeastern side of this, just where it sweeps round toward its northern terminus. The number of species which find their eastern or southeastern or southern limit with us will furnish something of a test of our relation to that region. After a careful review it is found that at least fifteen species having a western and southwestern distribution find their limit within our territory. Among these are *Jeffersonia, Gymnocladus, Solidago Ohioensis, Carya sulcata.* That some of these are found farther east, as near Oneida L., does not interfere with the general truth stated, viz., that they do not pass beyond us into Susquehanna basin so far as known. There are thirteen species having a decidedly northwestern, northern or northeastern range which have their southern terminus here.

These are:

Goodyera Menziesii,
Draba arabisans,
Saxifraga aizoides,
Lonicera oblongifolia,
Petasites palmata (probably),
Pinguicula vulgaris,
Primula Mistassinica,

Populus balsamifera,
Juncus alpinus, var. insignis,
Eleocharis pauciflora,
Carex gynocrates, (at Savannah),
   "   capillaris,
   "   Œderi.

The following apparently have their limit of distribution here but probably extend into the mountainous parts of Pennsylvania.

Viola renifolia,
Valeriana sylvatica,
Spiranthes Romanzoffiana,
Carex Deweyana,

Carex longirostris,
   "   pauciflora,
Cinna pendula,
Poa debilis.

The very remarkable sub-arctic group,—*Primula* and others,—may have been driven down here by the ice-sheet and have retained their foothold after its recession, finally retreating to the shaded wet walls of the ravines which were then forming, where they now remain, isolated from the home of the species. Several species are curiously modified in form and appearance, so as to resemble—according to a well-known biological law,—related species whose center of distribution is in this region. Such is the case with *Carya sulcata, Viola renifolia* and *Juncus alpinus*, var. *insignis.*

There are also visible traces in our flora of a connection with the Alleghanian plants. The best evidence is in the characteristic, common plants of the dry woods and ridges enumerated under the Lesser Floras. Indeed to any one familiar with the mountains of Virginia the correspondence is most striking. A few rare plants of Alleghanian origin probably find their northwestern limit with or but little north of us. These are *Dicentra eximia, Magnolia acuminata, Prenanthes serpentaria, Deyeuxia Porteri* and *Rhododendron maximum. Hydrangea arborescens*, found near Wellsburg on the Chemung R. and even in Warsaw Glen by Dr. Jordan, also by Miss Ross, will probably be found in our flora. There are a considerable number of species with a general southern range which disappear before they reach our northern limits. I have enumerated fifteen such, the most interesting of which are *Negundo aceroides, Agrimonia parviflora, Coreopsis discoidea, Polymnia Uvedalia, Potamogeton obtusifolius, Juncus marginatus* and *Cyperus Engelmanni.* That the Appalachian system has long interposed itself between us and the coast is evidenced by the few species peculiar to that region which terminate within our limits. *Solidago neglecta, var. linoides,* and *Andromeda ligustrina* with possibly *Carex glaucodea* are the only indigenous species recognized as belonging to that class.

It is believed that the following local species have not been reported from elsewhere in New York : *Goodyera Menziesii, Carya sulcata. Potamogeton obtusifolius, P. Hillii, Carex capillaris, Deyeuxia Porteri, Panicum nervosum, Polygonum nodosum.*

That the basin of Cayuga Lake was originally densely forested over three-fourths of its area, there can be little doubt. It also seems clear that the Cayuga tribe of Indians who were either occupants or overlords of all the territory within, and far south of our limits, had many cleared fields at the time of the arrival on the shores of our lake, of the Jesuits, Father Joseph Chaumonot and Father René Ménard, in Aug. 1656. Although they dwelt among the Indians until the remarkable flight of all the missionaries in Mar. 1657, before the supposed conspiracy of the League; and although they wrote voluminous letters of their life, their trials, their hopes and their failures, there is scarcely a word upon the aspect of the natural world which surrounded them. The mission at Cayuga was restored in 1668, and Stephen de Carheil remained there till 1684, when he was driven out by two Cayuga chiefs. Still we should know nothing of the region had not a Jesuit, Father Raffeix, who evidently had an observant mind, taken de Carheil's place, during a temporary absence of the latter. He was a man of wide experience and had visited the other tribes of the League in their own homes. He writes in the *Relations* for the year 1671–72, (Quebec Ed. p. 22): "Cayuga is the most beautiful country I have seen in America. It is situated in latitude 42½, and the needle dips scarcely more than ten degrees. It is a country situated between two lakes[1] and is no more than four leagues wide, with almost continuous plains bordered by beautiful forests. Agnie, (the country of the Mohawks), is a valley very narrow, often very stony, and always covered with fog; the hills which enclose it seems to me to be very poor land. Oneida and Onondaga, as well as Seneca, appear too rough and too little adapted to the chase. Every year in the vicinity of Cayuga more than a thousand deer are killed. Four leagues distant from here[2], on the brink of the river (Seneca outlet) are eight or ten fine salt fountains, in a small space. It is there that numbers of nets are spread for pigeons, and from seven to eight hundred are often taken at a single stroke of the net. Lake Tiohero, one of the two which join our canton, is fully fourteen leagues long and one or two broad. It abounds in swan and geese all winter, and in the spring one sees a continuous cloud of all sorts of game. The river which rises in the lake, soon divides into different channels enclosed by prairies, with here and there fine and attractive bays of considerable extent, excellent places for hunting." It is not difficult to picture to one's self the country here described. The marshes were as they are now, while all the country about, used by the Indians for the purpose of deer-stalking, was made up of "continuous plains" (the "oak-openings") bordered by the forest. The openings were kept clear by the Indians, by annually burning them over. These openings were described

---

[1] Owasco and Cayuga Lakes.

[2] He wrote from *Goiogoüen* or Cayuga, a town on the bank of Big Gully, and refers here to the salt springs north of the old Demont tavern and bridge, west side of Cayuga marshes.

by President Dwight in his tour through Western New York, 1822.[1]
Greenhalgh in his Journey "from Albany to ye Indians westward,"
1677, says of the Cayugas: "they have abundance of corne," which
implies of course cleared fields near their villages, at the foot of Cayuga Lake. But the universal testimony is that the forests became very
much denser and more tangled near the head of the lake and throughout the country south. The Cayugas made frequent excursions
through this country southward to the Susquehanna, and they are
spoken of in Hiawatha's decrees as the people whose "habitation was
the dark forest," their country being much more densely wooded
than Onondaga. John Bartram, a Pennsylvania Quaker and most excellent botanist, a keen observer and the most delightful of letter-
writers, made a journey to Onondaga in 1743, passing up the Susquehanna to Owego Creek. He then passed through the forest probably
traversing the east part of Tompkins Co., and the western part of
Cortland Co. He at first passed "over fine, level, rich land" with
"oak, birch, beech, ash, spruce, linden, elm, hepatica, and maiden-
hair in abundance." Then he struck "swampy land, then thickets,
and on the hills, spruce and white pine." Reaching level ground—
perhaps near the present site of Cortland—he found it "full of tall timber of sugar-maple, birch, linden, ash, beech, and shrubs of opulus,
green-maple, hornbeam, hamamelis, Solanum, gooseberries and red
currants." He describes the tops of the trees as so thick and inter-
lacing that it is "impossible to see which way the wind drives or the
clouds set." He next reached the dividing ridge where he found
chestnut and cherry in addition to the other trees; and toward Onondaga they found many "oaks, hickories, plums and apple-trees full of
fruit."

This corresponds well with the present distribution of species, so
far as the hardwoods are concerned. From Dr. Parker and other resi-
dents of Ithaca, whose recollections reach back to 1830 or 1835, we
know that the tracts of white-pine now wholly cleared away, were extensive and well-defined. There was a heavy growth of it in Enfield
stretching up for a long distance between the lakes. It descended into the Negueena valley near Buttermilk Falls, covered the top of
South Hill, extended back in several well defined tracts through Danby and Newfield and occupied portions of Negueena and other valleys. The pine was particularly heavy about Summit Marsh. There
were also large tracts in the upper part of Cascadilla valley, on Turkey Hill, and the pine-land extended north in narrow belts from
this region, penetrating the tracts of elm and maple of that region.
About Ithaca on the hills were "openings" of oak and hickory and
on both shores of the lake, these with other hardwoods prevailed.
In the valley near Ithaca, near Buttermilk Falls, and on both shores
of the lake there were apple-orchards, and cleared fields, cultivated by the Indians. On the bank near the Fleming school-
house was an Indian town called Coreorgauel, destroyed by Sullivan's army in 1779, and about the site Indian apple-trees existed

---

[1] Dwight's Travels, 1822, Vol. IV, p. 58.

dowy to a comparatively recent date. Although the Moravian missionaries, and particularly Bishop David Zeisberger, traversed this valley not infrequently, as early as 1750, it was not settled by white men
till 1789, so that all the changes in vegetation which have taken place
have come about in less than a century.

Sketch of the Explorations.

The first professional botanist to penetrate any portion of this basin
was John Bartram, whose journey to Onondaga in 1743, was described
above. His journey was not primarily a botanical one, and his book
entitled: "Observations on the Inhabitants etc., in Travels from Pennsylvania to Lake Ontario," was printed in London, 1751, and is now a
rare work.

Not many years after Peter Kalm, Professor of Œconomy at the
University of Aobo in Swedish Finland, a friend of Linnæus, was sent
over by the Swedish government, to collect seeds and plants for the
gardens and herbaria of that country. He collected extensively in
Pennsylvania, New Jersey, New York and Canada from 1748 till his
return in 1751. During 1750 he made a journey "along the Mohawk
to the Iroquois nations, where he got acquainted with the Mohawks,
Oneidas, Tuscaroras, Onondagas and Kayugaws." His "Travels into
North America," were published soon after his return and translated
into English in 1772; but while he gives minute accounts of some of
his experiences, his journey to, and life among the Six Nations is
scarcely touched upon in these volumes. His collections which were
very valuable went very largely into the Linnæan Herbarium, now in
England.

Frederick Pursh, another European explorer, passed through Ithaca in 1807. He left a record of his journey, which was over-
looked until 1868, when it came to light in Philadelphia among some
papers in the possession of the Amer. Phil. Society. It was published
in the Gardener's Monthly and afterward in pamphlet form. The
following literal extract from Pursh's *Journal,* is of some general interest, but chiefly so in revealing the abundance of introduced plants
even at that early day. He found nothing of importance in our region. His nomenclature will seem at the present time quite anti-
quated, and his English is somewhat broken :

"July 6.—[1807.]—Left Tyoga, up Cayuta Creek—Apocynum
andrasæmifolium very plenty in the cleared lands ; in the oak woods
I observed the Lily often seen before, but I cannot reccollect which
species it is ; it is Lilium foliis sparsis verticillatisque : caulle unifloro ;
floribus erectis, semipatentibus ; petalis unguiculatis. The valley,
formed by Cayuta Creek is in soil and vegetation similar to the beechwood. Oxalis acetosella—but not plenty—Helleborus viridis. Dra-
cæna borealis,—Orchis fimbriata in full bloom—Dalibarda violoides in
fruit. The woods about with sugar-maple ; The valley is in some
places very narrow and the creek very winding which obliged me
to wate it several times to keep the road—I heartyly expected to
reach the house this night, which had been recommended to me, to
stay at, but I dit come to it before it got dark. I observed in a small
run a species of Sium, as I suppose, without flowres, whose leaves under the water were very fine divided and the upper ones only pinnate. I call it S. heterophyllum. From a small tavern wheeh is kept

here, is about 22 miles to the head of Cayuga, which I intend to reach to-morrow.

7.—Having opportunity of going in company of a wagon, who would carry my things, I set out early this morning. The road leads through a very romantick valley, the mountains sometimes very high. After following the course of Cayuta Cr. for 9 miles, we turned off to the right. 8 miles this side of Cayuga city, or as it is called sometimes, Ithaca, we crossed a place very beautifully situated, called Sapony[1] Hollow ; this place has been once cleared and probably settled by Indians, but it is now grown up with small, white pine very handsomely mixed with Populus tremuloïdes and Magnolia acuminata. The last is very scarce about here and the trees here in this place and two or three others I seen are of a creeply, small and old growth, nothing like to what they are in Virginia. At this place we refreshed ourselves and feed the horses, as far as this I had this day travel very agreeable, as on account of the roughness of the road and the deep mire holes in some places the wagon could not go on as fast as I could walk, having plenty of time to look about myself, besides being unincumbered with any baggage. But the road getting now good and evening drawing nigh, I had to get into the wagon and we travele l tolerabiy fast. About 3 or 4 miles from Sapony Hollow, the timber changes into oak, and from there to Ithaca it is all Oak timber mixed with pine, with the rest of plants similar to Tyoga point. We arrived at Ithaca at night-fall.

8.—Being now on the heath of Cayuga I remembered your information about Erica cærulea growing on the high lands between Cayuga and Seneca lake ; I was very anxious of seeing this plant in its native place, but having not received the particular directions to find the place, as I had been promised of, besides that, being rather afraid of running myself out of money necessary to come to Onondaga, as my pocket was low and the distance considerable, I had to my own mortification, to give up all Ideas of a search for it.[2] The morning was rather suspicious for rain, as it had rained some all night, I was detained at Ithaca until 11 o'clock, when I set out for the lake, which is only two miles distance. My route was going on the east side of it. After having crossed Cayuga Creek, with a great deal of difficulty to perform it, and coming on the rising grounds, on the other side, I heard a very strong noise of falling water : I followed the sound and came to one of the most romantick and beautiful falls of this creek, I had ever seen ; the access even only to a sight of it is very difficult ; but regretted very much that I had not had the least information about these falls at the town, as I should have made it my business to visit them unincumbered with my baggage, that I might have spent the day by it. The ledge of rocks confined in a very narrow cove, and surrounded by very high hills ; impossible to ascend with a load on my back on account of steepness ; over which this considerable stream drops itself down, is a very interesting scene, and I doubt not if time and opportunity had allowed me to make an examination of it, I might have been paid for the trouble with something or other interesting or new in my line ; but to go back to the town I thought to be too much ; so I had to go on and be satisfied with having had only a peep at it. I got into my road again, where I

---

[1] Now called " Pony Hollow." Sapony Hollow is the only nominal relic in this region,—their last home,—of a once large tribe of Indians called Catawbas or Saponies, formerly residing in Virginia and North Carolina. There is indirect evidence that this last remnant sought protection of the Cayugas, and settled in this valley about 1753. Coreorganel was their principal town, and their burial places were near that village, also north of Buttermilk Falls and on the bluff near Dr. Parker's, East Hill, Ithaca. As a nation they were utterly extinguished by Sullivan's army as it swept over them, Sept., 1779.

[2] The supposed occurrence of Erica, (or Bryanthus taxifolius,) in this region, was, of course, a mistake.

observed along the banks of the creek plants of Pentstemon pubescens, About a mile farther I came to the bank of the lake. The shore which I came to was clear and gravelly with some common weeds growing near it as thistles, mulleins, etc. I followed the shore of it for several miles, being in my route. It is generally covered with oak, maple and hickory Buphthalmum heleanthoides is the first yellow Syngenesia plant I seen this year, Taraxacum excepted. A A small Rose, similar to the one I called last year R. monticola, is very plenty here and spreads a most agreeable fragrance through the air. A species of Cratægus, Ludwigia nitida, Ceanothus Americanus. Lilium Canadense, Apocynum androsæmifolium, with a tall Molugo? Orchis fimbriata, Cornus with white berries, Erigeron corymbosum, Typha angustifolia, Smyrnium cordatum, Mimulus alatus, Galium hispidum, Veronica scutellata and some more common plants, I observed in the meadows leading to the lake. I traveled as far as the town of Milton, where I stood over night. The road, as soon as I had left the banks of the lake began to be quit of interest, as the fences of both sides and cultivated fields, with continued plantations and farms occasioned the road only to be covered with common weeds, amongst which the Verbascum thapsus, Anthemis cotula, and Polygonum hydropiper, have the upper hand. In one of the woods on this road I collected specimens of Niphrodium filix-mas? [1]

David Thomas, who came from Pennsylvania in 1805, and settled in Cayuga Co. near Aurora, was at first a teacher, and afterward the engineer of the western division of the Erie Canal during its construction. He had an extensive botanic garden at his place east of Levanna and was an enthusiastic botanist as well as cultivator of plants. He was the first to distinguish and describe *Ulmus racemosa*, which he did in the Amer. Jour. of Sci. Vol. XIX, p. 170. The plates of this and those of the Dicentras in Vol. XXVI, were drawn by his son John J. Thomas, who made an extensive collection of the local plants in 1827, when he was seventeen. This herbarium is still well preserved and is frequently referred to in this catalogue. Some of the specimens are very valuable, in showing the character of the ealier flora. For the great kindness of Professor Thomas in going over with this herbarium in the writer's presence and stating many interesting facts, the latter wishes here to make due acknowledgement.

Dr. Aikin, a young medical student and pupil of Professor Amos Eaton, visited this region about 1830, and reported several rare plants. Dr. Jedediah Smith of Geneva, Dr. Alex. Thompson of Aurora, Dr, H. P. Sartwell of Penn Yan and Dr. S. B. Bradley of Greece all botanized more or less within our limits, and discovered many interesting things. Dr. Asa Gray, in 1831, visited the Montezuma Marshes and this lake, stopping at Sheldrake, and finding two rare plants *Pogonia pendula* aud *Linaria Elatine*. In Seneca Co., probably, he obtained *Goodyera Menziesii*. He visited Ithaca but did not remain long. Rev. J. W. Chickering and Prof. W. H. Brewer collected several hundred specimens between Cayuga and Seneca Lakes, while they were teachers at Ovid.

From 1860 to 1865, the Hon. H. B. Lord reported a considerable

---

[1] As Pursh elsewhere refers to Nephrodium marginale, the nearest relative of N. filix-mas, no doubt the one above mentioned was Aspidium Goldianum, a species not then described, but which is in his herbarium under the name of N. filix-mas, from an unknown American locality.

number of interesting plants and having been a resident of Ludlow-ville and Ithaca for many years he has given the younger generation of botanists very material aid in regard to localities and rare plants. He was particularly interested in *Carices* and gave to the University a valuable local collection, at the time of its opening. His name often occurs in this catalogue, but not nearly so often as his intimate and accurate knowledge of our plants would deserve.

At the opening of Cornell University in 1869, Professor A. N. Prentiss of Michigan Agr. Coll. was made Professor of Botany and Horticulture, and in spite of his administrative and other duties, has always manifested great interest in the development of the local flora. Under his careful guidance several special students, among them Mr. Theo. B. Comstock and Mr. David S. Jordan, both occupying prominent college positions at present, became deeply interested in the explorations and the latter summarized during his Senior year (1872), the results of his own experience and that of his friends in a manuscript catalogue. This was compiled from memory, and largely without the aid of preserved specimens and therefore contained some errors. But after throwing these out it still shows about 650 species. In this catalogue were 33 species of Carices, 35 species of Grasses, 6 Willows and 79 Compositæ.

There was at that time a group of men in the University who were strongly interested in botany and who never ceased to keep up that interest. These were Messrs H. E. Copeland, W. A. Kellermann, J. C. Branner, and the writer. Soon after these men left college, there came an excellect botanical student, Mr. F. B. Hine, whose name is mentioned frequently in the following pages. Then came Messrs. William Trelease, F. H. Severance and Charles Atwood, who added to our knowledge of the flora. Especially associated with my own work of exploration in 1881 and 1882, Mr. F. Cooper Curtice, now of the U. S. Geological Survey, rendered most efficient aid by his excellent observing powers.

The names of Mrs. Professor Brun, Miss I. Howland, Mr. F. L. Kilborne ; and among recent students, Mr. O. E. Pearce, Messrs. A. L. and F. V. Coville will be found after their discoveries in the following catalogue indicating in a slight way the aid they have rendered in voluntarily reporting localities or specimens.

Although the writer compiled a manuscript catalogue in 1876, containing 950 flowering plants, and has written out special catalogues of the Compositæ, Gramineæ and Vascular Cryptogamia, since that time his work on the flora for the past five years has been more systematic than before. He has kept slip records of the occurrence of every plant noticed on the numerous excursions made, providing there was any reason for supposing the plant was in the least uncommon. In some cases, as in some of the orchids and sedges these separate entries for distinct localities have been fifty or seventy-five, in other cases comparatively few. That the number of discoveries of species alone, new to the flora, has been 510 more than the number in Dr. Jordan's catalogue is in itself significant of the industry of the survey.

Statistics of the Catalogue.

The following tables show at a glance certain general as well as special facts in regard to the composition of our flora as it is set forth in the following catalogue. As the numbers in the catalogue are prefixed to both species and varieties, and also to a few forms unnamed, but which will probably be regarded as varieties when better known, and as some of the more important of recent catalogues use the same system of numbering, the following summaries will be based on that system and will include both forms. In certain important analyses, however, the species and varieties will be distinguished from one another.

It will be noticed that the catalogue includes several groups of specific and varietal names; these may be tabulated as follows:

  I. Species native to the Cayuga Flora 963
  II. Species introduced in the Cayuga Flora 197

    Total number of species 1160
  III. Varieties — native and introduced 118
    Total of numbered species and varieties 1278
  IV. Species spontaneous but not established 53
  V. Species in Seneca and Keuka Flora not in the Cayuga 42

No attempt has been made to enumerate the plants we have in excess of those known in the Seneca and Keuka region. That could be done by counting up the species of the catalogue which are followed by " H " or " C," or both ; but the comparison would not be wholly fair until after a systematic exploration of that region, especially in certain districts which have been quite neglected.

The following tables include only the numbered forms of the catalogue,—genera, species and varieties,—and have nothing to do with those names without numbers.

SYSTEMATIC DISTRIBUTION.

	Genera.	Species and Varieties.
Polypetalæ	162	381
Gamopetalæ	146	350
Total Dichlamydeæ	308	731
Apetalæ, (Monochlamydeæ) . .	45	133
Total Dicotyledons	353	864
Monocotyledons	101	403
Gymnospermæ	8	11
Total Phænogamia	462	1278

COMPOSITION OF THE PRINCIPAL ORDERS.

These are numbered according to the aggregate of *species and varieties* in each.

Orders.	Genera.	Species.	Species and Varieties.
1 Cyperaceæ . . .	9	120	151
2 Compositæ . . .	39	112	125
3 Gramineæ	44	93	107
4 Rosaceæ	17	57	69
5 Leguminosæ. . . .	17	42	45
6 Ranunculaceæ . .	12	34	36
7 Orchidaceæ . . .	13	35	35
8 Ericaceæ : . . .	16	30	35
9 Cruciferæ	14	31	34
10 Naiadaceæ . . .	20	25	34
11 Labiatæ	5	30	33
12 Scrophulariaceæ .	14	29	29
13 Polygonaceæ . .	2	26	28
14 Salicaceæ	2	26	28
15 Liliaceæ	17	27	27
16 Umbelliferæ. . . .	17	22	24
17 Caprifoliaceæ . .	7	20	22

THE PRINCIPAL GENERA.

The following are arranged according to the number of *species and varieties* in each.

Genera.	Species.	Species and Varieties.
1 Carex	84	112
3 Potamogeton	20	27
3 Aster	18	24
4 Salix	16	22
5 Polygonum	19	20
6 Solidago	16	18
7 Panicum	13	17
8 Juncus	11	15
9 Ranunculus	14	14
10 Viola	12	14
11 Scirpus	11	13
12 Habenaria	11	11
13 Prunus	10	10
14 Galium	8	10

In *Carex* and *Salix*, the few hybrids occurring are classed provisionally with the varieties.

COMPARISON WITH OTHER FLORAS.

It will be instructive to compare the results of our explorations with those exhibited in other, similar catalogues. For this we have selected several catalogues of certain representative sections of the State of New York, and a few well-known catalogues of more distant regions, where the work has been measurably thorough. Those selected from N. Y. are: 1. "*Plants of Oneida Co. and Vicinity;*" by John A. Paine, Jr., 1864; printed in the 18th Annual Report on the State Cabinet. This was a valuable catalogue, but covered in reality nearly the whole State, excepting the southeastern and the Adirondack regions. 2. "*Plants of Buffalo and Vicinity;*" by the Buff. Soc. of Nat. Hist., 1882. It includes the plants within a radius of thirty miles of Buffalo. 3. "*Plants growing without cultivation within five miles*

of Pine Plains, Dutchess Co., N. Y.;" by Dr. L. H. Hoysradt, 1875.
4. *"Plants of Suffolk Co., L. I.;"* by E. S. Miller and H. W. Young,
1874. Of these, the most suitable for comparison are the last three,
as they represent definitely outlined regions; the Buffalo list and the
Suffolk Co. list covering, of course, territorsies having very marked
characters. The other catalogues selected are Ward's *"Flora of
Washington and vicinity,"* 1881; Robinson's *"Flora of Essex Co.,
Mass.,"* 1880; and the *"Catalogue of Plants growing within thirty
miles of Yale College,"* 1878. In all cases the total numbers given
below refer to Phænogamia only. In comparing our own list with
the others mentioned, it may be safely said that the comparison is as
fair as such things can be in the present state of our knowledge. The
area covered—about sixty-five miles in length by an average of
eighteen in breadth—is equal to the average of those in table below.
There are no true mountains within the limits of any, although the
interesting little Flora of Pine Plains includes some of the spurs of
the Taconic range. On the other hand, that Flora has no true saline or shore plants such as most of the others possess. It has, however, a curious mixture of the plants of the cold marshes and moun-
tain-slopes of the north and those of the Atlantic coast.

Catalogue.	Total species and varieties.
The Cayuga Flora, ...	1278
Plants of Buffalo, etc., (with addend.)	1277
Plants of Oneida Co., etc.	1390
Plants of Dutchess Co.	1067
Plants of Suffolk Co.	852
Flora of Washington, etc.	1211
Flora of Essex Co.	1257
Flora of Vicinity of Yale College, . .	1238

It should be stated in regard to the latter list, that it does not include among its numbers the varieties mentioned in the body of the
work. Although a greater or less number of new discoveries have
been added to most of the above lists since their publication, no account is taken of such, excepting in the case of Buffalo Catalogue,
the facts not being accessible.

It will also be interesting to compare two of the larger genera, two
of the representative Atlantic coast genera and a representative northern genus, in respect to the abundance of species in different sections.

Genus.	Floras with the number of species and varieties in each.					
	Cayuga.	Buffalo.	Oneida Co.	Pine Plains.	Suffolk Co.	Washington.
Carex, ...	112	80	130	102	33	70
Aster, ...	24	24	29	20	20	21
Eupatorium, ...	4	3	6	4	10	12
Utricularia, ...	4	3	6	3	9	2
Habenaria, ...	12	10	16	9	3	4
(incl. Perularia.)						

The following is a comparison of the larger Orders :

Order.	E. U. S.	Cayuga.	Buffalo.	Oneida.	Pine Plains.	Suffolk.	Washington.
					No. Species and Varieties.		
Cyperaceæ, ...	357	151	115	187	138	79	108
Compositæ, ...	497	125	151	141	169	119	149
Gramineæ, . .	297	107	95	110	92		110
Rosaceæ, ...	104	69	55	62	48	35	46
Leguminosæ, ...	208	45	47	54	34	41	57
Ranunculaceæ, ...	80	36	39	36	27	17	27
Orchidaceæ, ...	71	35	35	42	30	15	24
Ericaceæ, ...	89	35	29	29	33	28	26
Cruciferæ, ...	76	34	37	32	27	32	34
Labiatæ, ...	121	33	43	40	27	31	42

The above estimates for the Eastern United States (E. U. S.) are taken from Ward's "*Flora of Washington.*"

The woo ly plants including woody vines, such as *Clematis* and *Menispermum*, and excluding, *Chimaphila*, *Gaultiera*, etc., are given below. Only so-called species are considered and are compared with the numbers in Sargent's Forest Trees of North America, Vol. IX, of the Tenth Census.

	Cayuga Flora.	The U. States.	Atlantic Region.
Total No. of species	206		
Native species.	175		
Introduced species.	31		
Native, arborescent species .	80	412	292
Introduced " "	18		

No attempt is made to compare the number of plants in our region with that of the whole state. Dr. Torrey's last estimate,—about 1853,— was 1537 species for the state of N. Y.; but, considering the large number of plants introduced since then, as well as the discoveries of native species, and knowing that we cannot always rely on mere printed reports, we do not think an accurate estimate can be made except by some one who is personally familiar with the different sections of N. Y.

Explanation of the Plan of the Catalogue.

1. Localities are given, so far as known, in case of "rare" or "scarce" plants.

2. Localities are occasionally given for species not rare. Such cases are indicated by adding the words, "and elsewhere."

3. The writer has in all cases endeavored to give proper credit in the catalogue to the discoverer of a new plant or new station of a scarce and rare plant whether the discoverer determined his plant or

not. All localities without name of person, are supposed to be from the personal observation of the writer. Wherever an exclamation point follows a name it signifies that the writer has verified the discovery, by himself observing the same station. Those plants so common as to need no special mention, are considered the common property of all observers, and therefore no discoverer's name is added.

4. Each species, variety, or especially marked form found with our Cayuga L. basin has its catalogue number, provided the plant is native, or has become established.

5. Those without numbers and without enclosing brackets are not regarded as yet permanently established within our limits.

6. Those without numbers but enclosed in brackets are found in the basins of Keuka or Seneca Lakes but not in that of Cayuga.

7. Species with names in heavy-faced type, are supposed to be indigenous.

8. Species with names in small capitals are not regarded as indigenous.

9. Whenever a species is not described in Gray's Manual 5th edition, a description is given in this catalogue if the form is regarded as permanent or important, and all references to Gray's Manual are to the 5th Edition.

10. The classification of Orders and Genera adopted, is that of Bentham and Hooker's *Genera Plantarum*; except that the arrangement in Gray's *Synoptical Flora* is preferred wherever that varies from the former. In the present unsettled condition of the *Carices* it was thought best to adhere to the old established arrangement of species as given by Carey in Gray's Man. The writer has also been conservative about adopting all the numerous changes made in the nomenclature of that genus.

11. The letters H., C., or H. and C., refer to the work of Dr. H. P. Sartwell, of Penn Yan, on the plants of the valleys of Seneca and Keuka Lakes. He was an accurate observer, an indefatigable worker, and in Senate Doc. No. 51, a report of the Regents of the University of the State of N. Y. for 1844, he published a catalogue of plants found in these two basins, and afterward disposed of his entire collection including specimens of most of those given in his catalogue, to Hamilton College. This collection 'is the " *Herb. Sartwell*" or the " H " of the following pages. "*Cat. of 1844*" or "C," refers to the above-named catalogue excluding those plants known to have been found only outside of the two lake basins. This catalogue contained 947 species,—not all of them found in the Seneca and Keuka regions—and to this Dr. S. H. Wright has for my convenience added all the recent discoveries known to him, 85 in number. His additions are indicated by the expression " Dr. Wright in C." Our cordial acknowledgements are due Dr. Wright for this and other kind acts. The following catalogue, therefore, furnishes not only the student of our flora a basis for future work, but the student of the Keuka and Seneca flora with similar, though less complete knowledge.

12. The (*) frequently used after the numbers indicate that the parti-

---

¹ A manuscript catalogue of this collection was obtained by the kind aid of Professor A. P. Kelsey, of Ham. Col.

cular species was given in Dr. Alex. Thompson's "List of Plants found in vicinity of Aurora" published in the Regents' Report for 1841. A certain number are rejected from that list as they were known not to have become really established. Another set inadvertently omitted in printing are given below, and should have the star in the text. These are :

Caltha palustris,	Asclepias tuberosa.
Rubus strigosus,	Gerardia flava,
" occidentalis,	" quercifolia,
" villosus,	Phryma Leptostachya,
" Canadensis,	Solanum Dulcamara,
Pirus Americana (introd.)	Rumex acetosella,
Anthemis Cotula.	Phytolacca decandra.
Tanacetum vulgare,	

13. In the expressions indicating rarity or abundance of a species, the word "rare" signifies that there are from one to four stations, and the specimens few in each. The other terms are applied with as much care as could be exercised, in the following order referring to their comparative abundance : scarce ; infrequent or uncommon ; not uncommon ; frequent ; abundant ; common.

14. Certain names of localities will strike many as unfamiliar. Old names have in several instances been revived, as in the case of the traditional Indian name—Neguæna—for that portion of the so-called Inlet above its junction with Six Mile Cr. Eagle Hill was the old name for what is often called "Bald Hill, Ithaca." But as there is another Bald Hill in Caroline, a revival of the old name seemed desirable. Since the opening of the University other names have been invented and adopted to meet the needs of collectors, and these are, of course, retained on the map and elsewhere.

The Disappearance of Species.

Certain species once known to exist within our limits have apparently disappeared. We do not find *Rhexia, Hippuris, Castilleia* or *Pogonia pendula ;* some of them may yet reappear, but probably not in the stations first reported. This catalogue reveals to all botanical students the stations of many rare plants, not before generally known. It is true that no student of our department, after he has come to have a real appreciation of his work, has ever needed the caution, to treat every plant as rare until it is known to be abundant, therefore, even this gentle reminder of a very good rule may not, perhaps, be strictly necessary. It may not, however, be out of place to point out the desirability of applying this rule to the plants of Cascadilla Woods, in particular. On account of its convenience to our laboratories, as well as its remarkable variety of forms, it is of almost priceless value, not only for its flowers, but the mosses and fungi it shelters. It is hoped, therefore, that not only will students be reasonably careful in collecting in it, but that all interested in the future of our campus will use every means to preserve the woods and its many interesting plants from any injury whatever.

CATALOGUE.

PHÆNOGAMIA.

DICOTYLEDONS.

RANUNCULACEÆ.

1. CLEMATIS, Linn.

1. **C. verticillaris**, DC. PURPLE CLEMATIS.   (C.)
   Cliffs and ravines; scarce. May 12–25.
   Cascadilla and Fall Creek ravines, a few stations. Buttermilk ravine below "Pulpit-Rock." McKinney's Twin-Glens. North of Esty-Glen. Taughannock, above the Falls.

2. **C. Virginiana**, L. WHITE C. VIRGIN'S BOWER.   (H. and C.)
   River-banks, thickets and swamps. July 25–Aug.
   Frequent near Ithaca, along the railroads, and throughout our lake-basin.

2. ANEMONE, Linn.

3. **A. cylindrica**, Gray.
   Roadside, Sherwood to Auburn, N. Y. June, 1880. (*Miss I. Howland.* See *Herb. Cornell Univ.*); rare.
   (Near outlet of Owasco L. *I. H. Hall in addend. of Paine's Cat.*)

4. * **A. Virginiana**, L.   (H. and C.)
   Wild banks, and lake shore; common. July.

5. **A. Virginiana**, L. var. **alba**, Wood, in Cascadilla Cr., etc.; frequent.

6. * **A. dichotoma**, L. (*A Pennsylvanica L.* Gray's Man., p. 37.)
                                                      (H. and C.)
   Abundant on gravelly shores of Cayuga Lake; also along the creek-beds and R. R. embankments. June–July.

7. * **A. nemorosa**, L. WIND-FLOWER.   (C. by Dr. S. H. Wright.)
   Copses and woods; frequent. May 1–20.
   Six Mile Creek, Negundo Woods, McGowan Woods, and elsewhere.

8. **A. nemorosa**, L. var. **quinquefolia**, Gr. occasional in Six Mile Cr. and elsewhere.

9. * **A. Hepatica**, L. HEPATICA. (*Hepatica triloba, Chaix*, Man. p. 38.)   (H. and C.)
   Dry woods of Pine, Oak, etc.; common. Mar. 25–May 15.
   Especially common in the dry woods on E. shore of Cayuga Lake. Specimens from South Hill, with stamens and pistils changed into sepals.

10. **A. acutiloba**, Lawson. HEPATICA. (*Hepatica acutiloba DC.* Man. p. 38.)   (H. and C.)

Richer soil than the preceding, in ravines and woods. Mar. 25-May 15.

Especially fine in Taughannock ravine, south slope. Woods about Freeville and the Romul Marshes. Not accompanying *A. Hepatica* in the dry woods on the east bank of the lake. Forms with five and even seven lobed leaves in rich shaded soil at Big Gully, Taughannock and other places. Specimens from Fall Cr. (*F. L. Kilborne*) have five enlarged divisions to the involucre (one being 3-lobed.)

3. ANEMONELLA, Spach.

11. *A. thalictroides, Spach. (*Thalictrum anemonoides* Mx. Man. p. 38. See Botanical Gaz., XI., p. 39, *A. Gray.*) RUE-ANEMONE. (H. and C.)

Rich or alluvial soil especially along ravines. Apr.-July.

Completely double flowers occasional in Cascadilla woods and elsewhere. Mrs. Brun finds such constant from year to year on plants from the same root. Specimens from six Mile Creek (1875) have all the flowers pale green. In other specimens an involucral leaflet has developed into a pure white sepaloid body.

4. THALICTRUM, Tourn.

12. *T. dioicum, L. EARLY MEADOW RUE. (H. and C.)

Rocky soil; frequent on the sides of ravines and on the lake shore. Apr. 20-May 15.

13. T. polygamum, Muhl. (*T. Cornuti L.* Man. p. 39) TALL MEADOW RUE. (H. and C.)

Swamps and meadows; frequent, especially around the springs of Mud Creek, Dryden Lake Valley and Michigan-Hollow. July, rarely earlier or later.

13ᵃ. T. purpurascens, L.

Scarce, in Fall Cr. and elsewhere.

5. RANUNCULUS, Linn.

14. R. aquatilis, L. var. trichophyllus, Ch. WHITE WATER-CROW-FOOT. (H. and C.)

In slow-flowing streams, or in still shallow water in the lakes; frequent. July.

Pools and bayous of Fall Creek and the marshes near. In the pockets of Cayuga L. cut off by the Cayuga So. R. R. Cayuga marshes, abundant. Clear pools in Beaver Cr. and upper part of Case. Cr., etc.

15. R. circinatus, Sibth. (*R. divaricatus Schrank.* Manual, p. 40); scarce.

Bayou in Fall Cr. marsh. Below Cayuga Bridge. "Marl Ponds" of South Cortland, where some of the plants are large and wholly submerged, others emersed, dwarf, 1-4 mm. high, leaves bright green, and flowers often with only four petals, blossoming till Oct.

16. R. multifidus, Pursh. YELLOW WATER-CROWFOOT. (H. and C.)

Ditches and overflowed swamps; scarce. Latter part of May. Lockwood's Flats, 1827. (*Herb. Professor J. J. Thomas.*) C. S.

R. R. ditches, from Glass-Works at Ithaca to the Corner-of-the-Lake. Along Cayuga St. near Corner-of-the-Lake. West Inlet-Marsh. Ringwood Swamp. Summit Marsh. (Penn Yan and Seneca L. *Sartwell.*)

17. R. **Cymbalaria**, Pursh. SEA-SIDE CROWFOOT.
Brackish marshes; local. June-Sept.
Montezuma, frequent E. of Salt Cr. and N. of the main road; observed in the same locality by Dr. Gray, 1831.

18. * R. **flammula**, L. var. **reptans**, Meyer. (H. and C.)
Wet, sandy shores; rare. July–Sept.
Cayuga Bridge, (*in Herb. Professor J.J. Thomas*). Shore on S.W. part of Myers Point, Cayuga Lake—found, Nov. 1880, (now in Univ. Herb.), but not seen since. Probably destroyed by the ice, which is often pressed up on the points and shores in winter, scoring up the loose sand. (Crooked Lake, *Sartwell.*)

[R. **ambigens**, Watson. (*R. alismaefolias, Geyer*, of Man. p. 41.)
"Hammondsport, Crooked L." *Sartwell in Herb. Ham. Coll.* Rood Swamp, near Painted Post, Steuben Co., *in Herb. from Miss I. S. Arnold*, 1884. It ought to be found within our limits.]

19. R. **abortivus**, L. SMALL-FLOWERED CROWFOOT. (H. and C.
Fields, meadows and damp woods; abund. May.

20. R. **sceleratus**, L. (H. and C.)
Ditches and wet places; not common. May to July.
Ithaca, abund. along the C. S. R. R. from the car-shops to Casc. Cr.; also beyond the Lehigh Valley depot. Union Springs and Canoga, occasional. Frequent on the Cayuga Marshes. (Seneca L. *Sartwell Herb. and Cat.* Also at Watkins!)

21. R. **recurvatus**, Poir. (H. and C.)
Ravines and damp woods; frequent. May–June.

22. R. **Pennsylvanicus**, L. (H. and C.)
Marshes, low grounds and wet places along streams; frequent. July–Sept.

23. R. **fascicularis**, Muhl. EARLY CROFOOT. (H. and C.)
Hillsides and thinly wooded slopes; frequent. Apr.–May.
Two forms occur here. In both, the earliest radical leaves are chiefly 3-parted, or 3-lobed. In the first (apparently typical) form the later radical leaves are ternate, each division twice pinnate with oblong or linear lobes; the cauline similar, the uppermost becoming simple. Roots fleshy and decidedly thickened. In the second form, the later radical leaves are ternate, the divisions coarsely serrate and wedge-shaped; the lateral divisions nearly sessile, usually lobed, the terminal stalked, broad and usually 2-parted; the cauline leaves 3-parted, 3-lobed, or the uppermost simple. Roots slender.

24. R. **repens**, L. (*R. Clintonii*, Beck, *Bot.*, p. 9, and Paine's Cat. p. 55.) (H. and C.)
The "low, hairy form," creeping and rooting at the nodes freely.

Rare ; at Union Springs in clay soil, near the Stone-Mill, appearing there as if introduced. May 20 June 20.

25. **R. septentrionalis**, Poir. (See *Pursh. Flora Amer. Sept.* II, p. 394 ; also Dr. Gray's *Revision of N. A. Ranunculi*, 1886.)
Marshy places ; common. May, June.
This form, referred to *R. repens*, L. since the publication of Torr. and Gray's Flora, is a much larger plant than that species, often nearly smooth and grows in wetter soils. The prostrate stems do not root freely, if at all.

26. R. BULBOSUS, L. BULBOUS CROWFOOT, (C.)
Near Big Gully, 1827. (*Herb. J. J. Thomas.*) Aurora, near R. R., and from the mouth of Paine's Creek to the road above the first falls ; apparently wanting in sections of our flora south of the above. (Near Geneva, 1885, *Fred V. Coville.*) A double-flowered *var.* occasionally escapes near gardens, as at McLean.

27. * R. ACRIS, L. BUTTERCUP. (H. and C.)
Fields ; very common. May–Oct.

6. CALTHA, Linn.

28. **C. palustris**, L. MARSH MARIGOLD. (H and C.)
Swamps and open wet meadows. April, May.
Fall Cr. in Cold-spring Marsh. Casc. Cr. below Judd's Falls. Frequent in marshes about Ithaca, (*T. B. Comstock*)! and abundant about Freeville, Round-Marshes, and the swamps of Groton and Danby.

7. TROLLIUS, Linn.

29. **T. laxus**, Salisb. GLOBE-FLOWER.
Swamps ; not common. May–June.
Near Larch-meadow. Swamps near Freeville, Malloryville, Mud Creek, Beaver Creek, W. Dryden, Groton, Danby and Newfield.

8. COPTIS, Salisb.

30. * **C. trifolia**, Salisb. GOLDTHREAD. (H. and C.)
Cold, mossy woods and marshes ; frequent. May 10-25.
Woods near Eddy Pond, (*Professor W. C. Cleveland*, 1871)! Six Mile Cr. Abundant on the hummocks, and about the margins of all the sphagnum marshes in the flora, and in the hemlock woods.

9. AQUILEGIA, Tourn.

31. * **A. Canadensis**, L. COLUMBINE. (H. and C.)
Rocky banks, and rocks of glens and lake-shore. May, June.
Also in cultivated grass-field, top of Bald Hill, Caroline.

32. A. VULGARIS, L. GARDEN COLUMBINE.
Established in woods beyond Larch Meadow. June.
Also by road on Dryden-Lansing town line ; Newfield in woods W. of R. R. sta., (*C. Humphrey in herb.* 1879) ; by road north of Cayuta Lake ; Saxon Hill, near Taber's.

DELPHINIUM, Tourn.

D. CONSOLIDA, L. LARKSPUR. (C.)
Six Mile Cr. 1875 (not established.)

10. HYDRASTIS, Linn.

33. **H. Canadensis,** L. GOLDEN SEAL. (H. and C.)
Woods and banks of ravines ; rare. May.
Six Mile Cr., near Sulphur Spr. 1872. (in flower May 24.) Swampy
woods head of Salmon Cr., N. E. of Merrifield's Sta. 1880. Big Gully
ravine (*in Herb. Mrs. Prof. Brun.*). Ledyard, (*Herb. J. J. Thomas.*)

11. ACTAEA, Linn.

34. **A. spicata,** L., var. rubra, Michx. RED COHOSH. (H. and C.)
Woods and wooded ravines ; frequent. Apr. 25–May 25.
Fall Cr., Six Mile Cr. and lake-shore ravines ; also wood in Dan-
by, and Bald Hill, Caroline. McGowan Woods. Dryden-Lansing
Sw., etc.

35. **A. alba,** Bigelow. WHITE COHOSH OR BANEBERRY.
(H. and C.)
Woods ; less frequent than above. May.
Cascadilla ravine. Fall Cr. Six Mile Cr. Freeville woods.
Bald Hill woods, etc.

12. CIMICIFUGA, Linn.

36. *** C. racemosa,** Ell. BLACK SNAKE ROOT. (H. and C.)
Wild grassy banks ; rare. Aug.
Ithaca, West Hill in Cliff-Park ravine. Lansing by road E. of
Taughannock Sta. E. side of Cayuta Lake. (E. side valley at
Watkins. Seneca L. Seneca, *Sartwell in Herb. and Cat.* Hector,
Herb. J. J. Thomas.)

[HELLEBORUS VIRIDIS, L. is reported from Penn Yan by Dr. S. H.
Wright.]

MAGNOLIACEÆ.

13. MAGNOLIA, Linn.

37. *** M. acuminata,** L. CUCUMBER TREE. (H. and C.)
Along ravines or in old woods ; not common. May 20–June 10.
Casc. Cr. above Judd's Falls. In McGowan Woods. Dart Woods
and Ringwood. Six Mile Cr., a large tree opposite the Sulphur
Spring. Along the valley to Brookton. Danby, in Durfee
Hill Woods abund.; also near Danby vill. and W. Danby. In Enfield and high hills of Newfield, frequent. Rare on W. Hill, Ithaca,
and in the low grounds about Freeville, McLean, etc., and only
occasional in the damp woods from Forest-Home north through
Groton, Genoa, to Sherwood and Big Gully Brook, and on the west
side of Cayuga L.

14. LIRIODENDRON, Linn.

38. *** L. tulipifera,** L. WHITE-WOOD. TULIP-TREE. (H. and C.)
With the preceding and somewhat more abundant. June 1–20.
Casc. Woods back of Armory and on banks of ravine. Fall Cr.
W. of Sibley Coll. Agrees with the preceding in scarcity or abund-
ance in certain districts. A tract of fine trees existed in 1882 ¼ mile
south of Big Gully.

MENISPERMACEÆ.

15. MENISPERMUM, Linn.

39. **M. Canadense**, L.　MOON-SEED VINE.　　　　(H. and C.)
Rich, rocky or alluvial soil.　June 20-30.
Case. ravine.　Fall Cr., Six Mile Cr. and other ravines.　Negundo Woods, and thickets near Fall Cr. on the marsh.　Freeville and McLean.

BERBERIDACEÆ.

16. BERBERIS, Linn.

40. *B. VULGARIS, L.　BARBERRY.　　　　　(H. and C.)
Pastures and fences; local and scarce.　May 25-June 15.
South Hill (*Dr. Jordan*), near the "Inclined Plane" of old R.R. !
By road to Buttermilk Falls.　Lane W. of Negundo Woods (*F. B. Hine*)!　Valley near "Rosedale."　Danby, on Michigan Hill.　By road N. of Forest-Home.　Ludlowville near Hedden and Townley Creeks.

17. CAULOPHYLLUM, Michx.

41. **C. thalictroides**, Michx.　BLUE-COHOSH.　　　(H. and C.)
Rich soil, ravines and woods; frequent.　Apr., May.
Fall Cr., Six Mile Cr., etc.　Freeville, McLean and Danby Woods.

18. JEFFERSONIA, Barton.

42. *J. diphylla, Pers.　TWIN-LEAF.
Rich soil, along shaded brooks; rare.　May 10-15.
Near Levanna.　(*J. J. Thomas in Herb. at Union Springs*), coll. in 1827 in woods near Barber's Brook and about a mile from the lake, apparently.
About two miles south-east of Levanna, near the east branch of Barber's Brook, in rich woods just west of the V. C. Deshong place; in flower May 14, 1882.
Near Aurora.　(*Dr. Alex. Thompson; See Report of the Regents of N. Y.*, 1841, p. 225.)
In Paine's Creek ravine, near Aurora; fine plants.　(*Prof. French of Wells Coll.*, 1883).
Near Wood's Station, L. A. and W. R. R., towards head of Salmon Cr.; discovered by *Dr. Charles Atwood, Moravia*, 1881.
Visited June 24, 1885, when the plants were found growing on a low, sandy, wooded bank north-east side of the pond east of the R. R.　They were luxuriant, covering eight or ten square feet, in fine fruit, and accompanied by the rare *Carex Careyana*, and by *C. tetanica*.
(Near Geneva.　*Sartwell, Herb. and Cat.*)

19. PODOPHYLLUM, Linn.

43. *P. peltatum, L.　MAY-APPLE.　MANDRAKE.　　(H and C.)
Rich soil; common.　May 20 June 10.
Specimens from Six Mile Cr., (*F. L. Kilborne*), 1883, show two distinct pistils, and many 3-lobed stamens.

NYMPHEACEÆ.

20. BRASENIA, Schreb.

44. *B. peltata, Pursh. WATER-SHIELD. (H. and C.)
Ponds and ditches ; rare. Aug.
Ithaca, ditch at corner of Fair Grounds, (*C. H. Wilmarth*, 1876);
also 1880! Locke Pond. Cayuta Lake, frequent. (Crooked L.,
Sartwell, Herb. and Cat.)

21. NYMPHÆA, Tourn.

45. *N. odorata, Ait. WHITE POND-LILY. (H. and C.)
Shallow still water ; not common. July–Aug.
Canoga marshes opposite Union Springs. Near Cayuga Bridge,
a few. Locke Pond. Cayuta Lake. (Penn Yan, *Sartwell, Herb.
and Cat.*)

46. N. odorata, Ait. var. minor, Sims.
Summit Marsh. Cayuta Lake.

47. N. tuberosa, Paine.
In the larger marshes. July, Aug.
From Cayuga Bridge, north ; abund. near Black Lake.

22. NUPHAR, Smith.

48. N. advena, Ait. YELLOW POND-LILY. (H. and C.)
Shallow water ; common. June, Aug.
Marshes about Ithaca, Cayuga Marshes, Locke Pond, Freeville, etc.

49. * N. pumilum, Smith, (*N. luteum, Sm. var. pumilum*, Man. p. 57.)
Bayou near mouth of Fall Cr., 1876. Ditches near Ithaca Fair-
Grounds. Groton Mill Pond, (*Prof. S. G. Williams*, 1877) ; same
station 1881! (North Spencer Pond, 1880. Flint Cr., Gorham,
Sartwell, Herb. and Cat.) Flowers have never been found in this
region.

SARRACENIACEÆ.

23. SARRACENIA, Tourn.

50. * S. purpurea, L. PITCHER-PLANT. (H. and C.)
Peat-bogs. June 10–20.
Larch-meadow, Ithaca, 1873, (*Dr. J. C. Branner,*) now nearly ex-
tinct. Freeville bog, 1872, (*Dr. Jordan.*) Frequent in this bog, at
Malloryville, Woodwardia Sw., and the Round-Marshes.
The *S. heterophylla* form at "W. Junius." (*Sartwell, Herb. and
Cat.*) W. Junius, near Lowery's Ponds and north of Pout Ponds!

PAPAVERACEÆ.

PAPAVER, Linn.

P. SOMNIFERUM, L. (COMMON POPPY.) By road S. E. of Edgewood,
1875 and 1881.

P. RHŒAS, L. (FIELD POPPY.) Campus near McGraw Building,
1875 and 1883. Fiske-McGraw grounds, 1883.

24. CHELIDONIUM, Linn.

51. C. MAJUS, L. CELANDINE. (H. and C.)
Shaded turfless places ; frequent. June.

Roadsides near dwellings. Also in woods beyond Buttermilk
Falls bridge, abund., and in the amphitheatre of Six Mile Cr. Near
Saxon Hill, etc.

25. SANGUINARIA, Dill.

52/ * **S. Canadensis**, L.   BLOOD-ROOT.   (H. and C.)
Frequent in Fall Cr. and other ravines, and in thickets. April
15–30.

FUMARIACEÆ.

26. ADLUMIA, Raf.

53. **A. cirrhosa**, Raf.   ALLEGHANY-VINE.
Ithaca, foot of Cascadilla ravine, 1873; not permanent.
"Junius." (*Sartwell, Herb.*) (In a rocky thicket, road between
Watkins and Havana, N. Y., June, 1884 ; apparently indigenous.)

27. DICENTRA, Bork.

54. * **D. Cucullaria**, DC.   DUTCHMAN'S BREECHES.   (H. and C.)
Common in Fall Cr. and the other ravines ; also in rich woods
near Aurora, Union Springs, Freeville and in Negundo Woods,
April 15–30.

55. **D. Canadensis**, DC.   SQUIRREL CORN.   (H. and C.)
Rich alluvial soil. Woods and ravines ; frequent. Apr. 20–30.
Fall Cr., Six Mile Cr., Negundo Woods, especially luxuriant.
Freeville. Danby. Paine's Cr.

56. **D. eximia**, DC.
Junius, Seneca Co., *in Sartwell's Herb.* (Probably only one sta-
tion ; *either Junius* or in *Wayne Co.*)
"Dr. Sartwell is the only botanist who has found it in the limits
of N. Y." (See Torrey, in Flora of N. Y., I, p. 47, 1843.)
In the 19th Report of the Regents, on the N. Y. State Cab. of
Nat. Hist., p. 77, (1865), Judge Clinton says, under *Dicentra eximia*,
DC.: "On recurring to my correspondence with David Thomas, in
1829, I find that he had not then found this plant native in Cayuga
Co. Prof. Pickett kindly communicated to me a letter of my dear
friend, Dr. Sartwell, dated June 23, 1865, in which he writes : '.As
to *Dicentra eximia*, I know not where it can be found. About
twenty years ago I found it in Wayne Co., not far from Lyons ; and
David Thomas found it in Scipio, Cayuga Co., about the same time
or before. * * I have no duplicate specimen.'" Gray, in Manual,
(1865), p. 61, gives 'Rocks West. N.Y.' (*Thomas, Sartwell.*) Paine's
statement (Cat., p. 59), that Sartwell found it near Sodus Bay, is
probably an error. But it is still uncertain whether it was originally
found in *Seneca* or *Wayne Co.*

28. CORYDALIS, Vent.

57. **C. glauca**, Pursh.
Cliffs, Fall Cr., north side below Forest-Home ; rare. May 15–30.
(*Discovered by C. O. Upton, May 14, 1878.*)!
"Junius," (*Sartwell, Herb. and Cat.*)

58. **C. aurea**, Wild. GOLDEN CORYDALIS.
Rocks below Lucifer Falls. June 1-10.
(*Discovered by Prof. A. N. Prentiss, June, 1871.*) Several times
it has seemed to have become extinct, but it reappears, and flowers
have been collected, 1875, 1882, 1884, 1885.

29. FUMARIA, Linn.

59. F. OFFICINALIS, L. FUMITORY.
McKinneys, by the C. S. R. R., a few rods south of the station.
June 3, 1882, in flower; also June 16, 1885; plants spreading.

CRUCIFERÆ.

30. LUNARIA, Linn.

60. L. BIENNIS, L. COMMON HONESTY.
Escaped; Dryden Road, by hedge south side of Giles Place,
May 20-30.
Fall Cr., near sand-bank above Forest-Home. (*E. H. Prestwick.*)

31. ALYSSUM, Tourn.

61. A. CALYCINUM, L.
Introduced: at the Pres. White place, Univ. Campus, May 20,
1878, (*Professor Prentiss*)! Along the road near the Jayne Place,
1879, (*Frank H. Severance*)! Fiske-McGraw grounds, 1881, (*F. L.
Kilborne.*)!

32. DRABA, Linn.

62. D. arabisans, Michx.
Margins of cliffs; rare—a northern species. May 20-June 10.
Burdick's Glen, along the brook just above the falls, 1871. First
small ravine below McKinney's Glens, Nov., 1881.

63. D. verna, L. WHITLOW-GRASS.
Shaded clayey soil; scarce. Apr.-May 15.
Introduced in Cemetery, Ithaca, (*1878, Prof. Prentiss*)! Wild,
south bank of Big Gully, 1882.

33. DENTARIA, Linn.

64. * D. diphylla, L. PEPPER-ROOT.   (H. and C.)
Ravines and rich woods; frequent. Apr. 20-May 20.
Casc. Cr. and all ravines, often in springy soil. Negundo, Free-
ville and Round Marsh Woods.

65. D. maxima, *Nutt.*
Bottoms of ravines; scarce. Apr. 20-May 20.
Six Mile Creek, Beechwoods, where specimens have been obtained
like those mentioned in Paine's Cat., viz.: with a secondary branch
arising from the axil of the third and upper leaf. Ravine on the
Renwick farm slope, and doubtful forms from Taughannock below
the falls.

66. D. laciniata, Muhlenberg.   (H. and C.
Ravines, rich woods; common and variable. Apr. 20-May 20.
The typical form, with leaves whorled and leaflets freely and
sharply toothed, occurs in Negundo Woods, Six Mile Cr., Fall Cr.,

Franklin's Ravine, Taughannock, etc. The form with alternate more or less long petioled leaves is with the other, and especially abundant at Negundo Woods. A form, tall—25-35 cm., with 2 leaves, each with 3 broad rather simple leaflets, and purplish flowers, but usually with the rootstocks of *D. laciniata*, is placed here by Dr. Gray. This is the best disposition to be made of the numerous forms of this abundant species, so far as its local specimens are concerned.

34. CARDAMINE, Linn.

67. **C. rhomboidea**, DC.   WH. SPRING-CRESS.   (H. and C.)
Rich woods or sometimes wet places.   May 10-30.
Indian Spring Marsh.   Near Freeville and McLean ; less common than the *var. purpurea*.

68. *C. rhomboidea**, var. purpurea, Torr.   PURPLE SPRING-CRESS.
In wet and springy places ; common.   April 20-May 15.
It blooms earlier, the aspect is different and like a distinct species.

69. **C. pratensis**, L.   CUCKOO-FLOWER.   (H. and C.
Wet marshes and meadows ; scarce.   May.
Marshes, head of Lake, and Dryden-Lansing Sw.   (*Jordan*)! Larch Meadow.   Fleming Meadow.   Near Freeville.   Round Marsh.   Locke Pond.

70. **C. hirsuta**, L.   (H. and C.)
Wet places, ravines and brooksides ; common.   May-July.

71. **C. hirsuta**, L., var. sylvatica, Gr.
Drier places ; rare.   June.
High rocks and shales, Burdick's Glen and vicinity.   (*F. L. Kilborne*, 1882)!

35. ARABIS, Linn.

72. **A. lyrata**, L.
Rocks and shale, ravines and cliffs ; scarce.   June, July.
Rare in Fall Cr. and Enfield ravines.   Caroline, somewhat abundant on the North Pinnacle ; almost wholly wanting on the lake-shore cliffs.   Cliffs S. E. corner of Seneca L.

73. **A. hirsuta**, Scop.   (C.)
Cliffs and rocks ; frequent.   May 20-June 10.
Fall Cr.   Enfield ravine.   Six Mile Cr., "Narrows."   Lake shore ravines.   In shaded places it produces small thin-leaved specimens, resembling *Sisymbrium Thaliana*.

74. **A. laevigata**, DC.   ROCK CRESS.   (H. and C.)
Rocky places ; frequent.   May 20-June.
Cascadilla, and other ravines.

75. **A. Canadensis**, L.   SICKLE-POD.   (H. and C.)
Ravines and hill slopes, in thickets widely distributed, but growing singly.   June-Aug.
Fall Cr.   Six Mile Cr.   Enfield ravine.   South Hill and lake-shore ravines.

76. **A. perfoliata**, Lam.

Fields, appearing as if introduced ; scarce. June–Aug.

South Hill, above the Quarry. Ulysses near Perry. Near Summit Marsh.

77. **A. Drummondii**, Gray.

Rocks and shale cliffs ; frequent. June.

Fall Cr. and all the ravines, especially abundant on the High Cliffs north of King's Ferry.

36. ERYSIMUM, Linn.

78. **E. cheiranthoides**, L.   (H. and **C**.)

Dry soil and thickets ; not common. June–Sept.

Near the Armory. Near the " Nook." East and west shores of the Lake.

37. BARBAREA, R. Br.

79. **B**. vulgaris, R. Br.   WINTER CRESS.   (H. and **C**.)

Cult. soil ; everywhere. May, June.

38. SISYMBRIUM, Linn.

80. S. OFFICINALE, Scop. HEDGE MUSTARD.   (H. and **C**.)

Waste places ; abundant. June–Sept.

81. **S. canescens**, Nutt. TANSY MUSTARD.

Cliffs ; very rare ; June.

" Lucifer Falls, Tompkins Co. J. W. Chickering," (*Gray's Man.*)

I have not been able to find it there, but it grows above the footpath at the entrance of Watkins Glen, (*Dr. Jordan discoverer.*) !

HESPERIS, Linn.

H. MATRONALIS, L. Scarce, probably not established. July.

Road toward Buttermilk Falls, coll. in 1875 and 1882. Near Giles Place.

39. BRASSICA, Tourn.

82. B. SINAPISTRUM, Boissier. YELLOW MUSTARD.   (H. and **C**.)

Cult.-fields, etc. June–Sept.

83. * B. NIGRA, Gray. BLACK MUSTARD.   (H. and **C**.)

Waste places and along streams and lake shores. June–Oct.

84. B. ALBA, Gray.

Fields ; rare. July, Aug.

In Ithaca on Seneca St. Field E. of Freeville.

B. OLERACEA. (CABBAGE.) Springs up in many places ; abundant in 1884.

40. NASTURTIUM, R. Br.

85. N. OFFICINALE, R. Br. WATER-CRESS.

Spring-brooks, etc. July–Aug.

Abund. in the ditches from the springs in Ithaca village. Indian Spring, etc.

86. **N. palustre**, DC. MARSH-CRESS.   (H. and **C**.)

Wet places, borders of pools, etc. June–Sept.

The typical form, nearly smooth and with oblong pods, is found in Cascadilla ravine, a few; and at the Marl Ponds of Cortland.

87. **N. palustre**, DC. var. **hispidum**, Gr.   (**H.** and **C.**)
   Common on marshes about Ithaca, Cayuga Marshes, Summit
   Marsh and about streams and pools elsewhere.

88. *N. ARMORACIA, Fries. HORSE-RADISH.   (**H.** and **C.**)
   Wet places and roadside ditches.   May, June.
   Thoroughly at home in the wet marshes at the Corner-of-the-Lake,
   where it blooms freely.   Also along the ditches south of town, and
   the village streets.

41. CAMELINA, Crantz.

89. C. SATIVA, Crantz. FALSE-FLAX.   (**H.** and **C.**)
   The University Campus, 1875, and 1882. June.
   Also near Aurora by the R. R. (Watkins, E. side of valley.
   "Penn Yan, frequent." *Sartw., in Herb.*)

42. CAPSELLA, Vent.

90. *C. BURSA-PASTORIS, Moench. SHEPHERD'S PURSE. (**H.** and **C.**)
   Everywhere.   Mar–Dec.

43. LEPIDIUM, Linn.

91. L. VIRGINICUM, L. PEPPER-GRASS.   (**H.** and **C.**)
   Dry waste grounds and roadsides; common. June–Oct.

92. L. RUDERALE, L. PEPPER-GRASS.
   With the preceding and equally common; distinguished from it
   by its somewhat smaller pods, and darker green color. Occasionally
   it has fasciated recemes.

93. L. CAMPESTRE, L.
   Fields and yards. June, July.
   First noticed in Ithaca, 1873. Near the "Nook." South Hill.
   Much more abundant near Union Springs and Aurora.

RAPHANUS, Linn.

R. SATIVUS, L. RADISH.
   Occasional scape ; not uncommon on the lake shore.

CAPPARIDACEÆ.

44. POLANISIA, Raf.

94. *P. graveoleus, Raf.   (**H.** and **C.**)
   Gravelly banks ; near the ravines, and on the lake shore; frequent. July.

CISTACEÆ.

45. HELIANTHEMUM, Tourn.

95. H. Canadense, Michx. FROST-WEED.
   Dry banks of ravines ; scarce. June 20–July.
   Near mouth of Buttermilk Glen. Coy Glen. Salmon Cr. ravine.
   Valley Cemetery south of Ithaca.   "Junius," (*Sartwell, H. and
   C.*); about Newton's ponds !

46. LECHEA, Linn.

96. L. thymifolia, Pursh. PINWEED.
   Dry soil ; rare. Summer. North side Salmon Cr. ravine.
   (L. Major, Michx. *Avon* and L. minor, Lam. *P. Yan*, "in
   cemeteries," *Sartwell, Herb. and C.*)

VIOLACEÆ.

47. IONIDIUM, Vent. (*Solea, Ging.* of Gray's Man.)

97. I. concolor, Benth and Hook. (*S. concolor* of Man., p. 77.) GREEN
VIOLET.   (H. and C.)

Slopes of ravines ; rare.   May 25-June 10.

"Woods, Ledyard," 1827, (*in Herb. J. J. Thomas.*)   Salmon
Creek ravine north side.   Six Mile Creek, east of Sulphur Spring,
in 1874 and 1876.

48. VIOLA, Linn.

98. V. rotundifolia, Michx.   SWEET YELLOW VIOLET.

Rich cold woods, where it is frequent.   Apr. 15-May 10.

Near Brookton (*Dr. Jordan,* 1871)!   Six Mile Cr. south of Nar-
rows, (*Branner,* 1873)!   Near Buttermilk ravine.   Enfield ravine.
Lick Brook, Dart Woods, Caroline Hills, Swamps of Dryden and
Danby,

99. V. blanda, L.   WHITE VIOLET.   (H. and C.)

Ravines, wet woods and swamps ; common.   May.

100. V. renifolia, Gray, (*Proc. of Amer. Acad. of Arts and Sci.,* VIII.,
p. 288.)

"Rhizome and flowers like V. blanda L. or the flowers a little
larger.   Leaves reniform usually 5 cm. broad, both sides and the
petioles villous-pubescent, scape pubescent."   Growing with *V.
blanda* and like it except its "reniform leaves are conspicuously
beset with pale, soft and tender, lax. hairs."   Deep mossy swamps,
like the Round-Marshes and Mich. Hollow Sw.; not uncommon.
The hairs are not as conspicuous, in our specimens, nor the scapes
or petioles as stout, as in northern specimens.   We are probably
near the southern limit of the species.

101. *V. ODORATA, L.   "ENGLISH VIOLET."   Sometimes spreads
beyonds the gardens, as in Ithaca, on Seneca St. near Spring St.
and elsewhere.

102. V. Selkirkii, Pursh.   (Goldie.)   GREAT-SPURRED VIOLET.

Shaded ravines or by rivulets ; rare.

"In Enfield near Lucifer Fall," (*W. J. Freeman, May* 15, 1880),
in Herb. of C. U.

103. *V. cucullata, Ait.   COMMON BLUE VIOLET.   (H. and C.)

Open thickets, meadows and swamps ; common.   May-June.

104. V. cucullata, Ait., var. palmata, Gray.   HAND-LEAVED V.
                                            (H. and C.)

In drier situations, grassy slopes ; frequent.

University grove.   North side of Fall Cr.   Renwick farm slope.
Woods and glens on both shores of Cayuga Lake.   Specimens from
the University grove, (*F. L. K., May,* 1878), show entirely regular,
spurless and beardless flowers.

105. V. sagittata, Ait.

Dry hillsides ; not common.   May.

Abundant on South Hill near N. Y. S. S. 420, and south, (1871).
Near Buttermilk ravine.   West Hill, near Cayuga Lake, (*Trelease,*)

1879. Near White Church. Danby, S. E. of village ; Thacher's Pinnacle and south. Hill north of Signer's Woods. Renwick.

106. **V. canina**, L. var. **sylvestris**, Regel. DOG VIOLET. (H. and C.)
Rich, damp woods and borders of marshes ; abund. May–June.
Fall Cr. and Six Mile Cr. Negundo Woods ; and especially abundant near Beaver Cr., and the Pond Meadow near.

107. **V. rostrata**, Pursh. LONG-SPURRED VIOLET. (H. and C.)
Ravines and rich woods ; common. May, June.
Casc. Cr., and all ravines. McGowan Woods. Woods of Dryden, Danby, Lansing and north. Specimens coll. May, 1882, show double flowers, the sepels, petals and stamens being 6–9 each, the number of spurred petals 2–4 in each flower, and the placenta 3 or rarely 4.

108. **V. striata**, Ait. PALE VIOLET.
Rich woods ; not common. May 15–June 15.
Fall Cr. near Cold Spring. Six Mile Cr. Negundo Woods. McGowan Woods. Taughannock ravine. Woods of Freeville and Round Marsh.

109. **V. Canadensis**, L. (H. and C.)
Ravines and rich swampy woods. May–July.

110. * **V. pubescens**, Ait. DOWNY YELLOW VIOLET. (H. and C.
Ravines and rich woods. May–June.

111. **V. pubescens**, Ait., var. **eriocarpa**, Nutt.
Mead Woods. McGowan Woods. Six Mile Cr. Renwick Farm. Freeville, abund.

POLYGALACEÆ.

49. POLYGALA, Tourn.

112. **P. verticillata**, L. (H. and C.)
Dry soil, on banks of the ravines and Cayuga L. shore. July–Sept.

113. **P. ambigua**, Nutt. (C.)
Similar situations ; scarce.
Big Gully, 1827, (*in Herb. J. J. Thomas.*)

[**P. sanguinea**, L., in Sartwell's Cat.]

114. * **P. Senega**, L. SENECA SNAKE-ROOT. (H. and C.)
Slopes of ravines ; frequent. May 15–June 15.
Casc. Woods. Six Mile Cr. Fall Cr. Ravines of Cayuga Lake. Bald Hill and the Pinnacles near White Church.

115. * **P. paucifolia**, Willd. FRINGED POLYGALA. INDIAN PINK. (H. and C.)
Thickets and along ravines ; widely distributed. May.

CARYOPHYLLACEÆ.

50. DIANTHUS, Linn.

116. D. ARMERIA, L.
Fields ; scarce. DEPTFORD-PINK. July–Aug.
Near the road south of Burdick's Glen. In Cascadilla Valley near Turkey Hill. Meadow S. W. of Freeville.

117. D. BARBATUS, L. SWEET WILLIAM.
    Established in woods by Rumsey's Brook, eastern base of Saxon
Hill. In Case. Woods, June, July, 1884 and 1885, (*Miss Hale.*)
Lot south of the cemetery, (*F. T. Wilson.*)

51. SAPONARIA, Linn.

118. *S. OFFICINALIS, L. BOUNCING BET.   (H. and C.)
    Roadsides, and near brooks. July–Sept.

52. SILENE, Linn.

119. **S. stellata**, Ait. STARRY CAMPION.   (H. and C.)
    Banks of ravines; scarce. July–Aug.
    Beyond the "Nook." Renwick Farm slope. Salmon Cr. ravine.

S. ARMERIA, L.
    Ithaca, by road south-east of steamboat landing, July, 1885. Near
Six Mile Cr. by Cayuga St.

120. **S. antirrhina**, L. SLEEPY CATCHFLY.   (H. and C.)
    Fields or dry banks; not common. May–June.
    Cascadilla Woods, abund. in 1885. By R. R. south of Ithaca.
The cedar knoll near High Cliffs of Cayuga L. Near McKinney's.

" **S. Pennsylvanica**, Michx.   (C.)
    "In vicinity of Aurora," 1840, in catalogue of Dr. Alex. Thomp-
son, as native, but probably a garden-scape.

[S. VIRGINICA, L. *Sartwell, in H. and C.* Penn Yan. Probably
not established.]

[S. NOCUTURNA, L. *Sartwell, in C.* Probably not established.]

121. S. NOCTIFLORA, L. NIGHT-FLOWERING CATCHFLY. (H. and C.)
    Waste places; frequent. June–Oct.

53. LYCHNIS, Tourn.

L. VESPERTINA, Sibth.
    Fiske-McGraw grounds, west of the spring, 1884 and 1885.

122. *L. GITHAGO, Lam. CORN COCKLE.   (H. and C.)
    Grain-fields; frequent. June.

54. CERASTIUM, Linn.

123. C. VISCOSUM, L. MOUSE-EAR CHICKWEED. (C. by Dr. Wright.)
    Fields; common. Mar.–Dec.

" C. VULGATUM, L. (*in Sartwell's Catalogue.*)

124. **C. nutans**, Raf.
    Moist places, in ravines or woods. May–June.
    Fall Cr. near mill-pond. Coy Glen. Renwick Farm. South
Hill. Lake-shore ravines. Swampy woods north of Freeville.
(Watkins, E. side of valley)!

125. **C. arvense**, L.   (C.)
    Rocky places; scarce.
    Ithaca cemetery, probably introduced; also Fiske-McGraw
grounds. (Watkins, E. side of valley on rocks; native.)

55. STELLARIA, Linn.

126. *S. MEDIA, Smith. CHICKWEED. (H. and C.)
Fields, etc. Mar.-Dec.

127. S. longifolia, Muhl. STITCHWORT. (H. and C.)
Wet places and swamps. May 15-July.
Indian Spr. Marsh. Formerly on Univ. Campus N. W. of the
Pres. White place, (F. C. Curtice,)! Freeville. Beaver Creek Swamp.
Brookton Spr.

128. S. borealis, Bigelow. NORTHERN-STITCHWORT. (H. and C.)
Moist soil ; scarce ; June.
South Hill. Swamps of W. Dryden, Freeville and Beaver Cr.

S. GRAMINEA. L.
Introduced in several places on the Fiske-McGraw grounds, 1885 !

56. ARENARIA, Linn.

129. A. SERPYLLIFOLIA, L. SANDWORT. (H. and C.)
Fields, gardens, etc. May-Dec.

130. A. lateriflora, L. (H. and C.)
Marshes and Sphagnum Swamps ; scarce. May 20-June.
Indian Spr. Marsh. Larch Meadow. Fleming Meadow. Across
the road from the Valley Cemetery. Freeville, in Mud Cr. Sw.
Malloryville Marsh. South Hill, Ithaca, near the quarry, (F. C. C.)
Lockwood's Flats, (Herb. J. J. Thomas.)

57. SPERGULA, Linn.

131. S. ARVENSIS, L. CORN-SPURRY. (C.)
Cult. fields ; at present only on the higher hills ; not common.
July-Aug.
Snyder Hill. Turkey Hill. Top of Taft Hill, Caroline. Top of
Saxon Hill, Newfield. High hills south of Danby.

PORTULACACEÆ.

58. PORTULACA, Tourn.

132. P. OLERACEA, L. PURSLANE. (H. and C.)
Gardens and roadsides. July-Sept.

59. CLAYTONIA, Linn.

133. *C. Virginica, L. SPRING BEAUTY. (H. and C.)
Rich woods ; abundant Apr. 15-May 15.
Six Mile Cr. Mead Woods. Negundo Woods. Freeville, McLean,
and especially abundant in pastures E. of Levanna and Union Springs.

134. C. Caroliniana, Michx. SPRING BEAUTY. (H. and C.)
Rich woods, often in shaded swamps, in moss ; rather less common than the preceding. Apr. 15-May 15.
Six Mile Cr. Negundo Woods, and about the same range as the
preceding. It mostly replaces it on the higher ground farther from
the lake, E. of Levanna and Union Springs.

HYPERICACEÆ.

60. Hypericum, Linn.

135. H. Ascyron, L. (*H. pyramidatum, Ait.*, Man., p. 84.) Tall St. Johnswort. (H. and C.)

Banks of streams, in rich soil; scarce. July 15–30.

Fall Cr. near Varna; (*Dr. Jordan*, 1871); Beebe Pond; near Cayuga St. bridge. Casc. Cr.; below Genung's Mill; above Judds Falls; head of Eddy Pond; in the ravine opposite Cascadilla Place.

136. H. perforatum, L. Common St. Johnswort. (H. and C.)

Fields; very abundant. July–Aug.

137. H. maculatum, Walter. (*H. corymbosum*, Muhl., Man., p. 85.) (H. and C.)

Banks and shaded places; frequent. July–Aug.

Near Willow Pond race. Fall Cr. Pastures and woods near Freeville, McLean, and elsewhere.

138. H. mutilum, L. (H. and C.)

Wet places; common, July–Sept.

An erect form on Goodwin's Point, Cayuga L.

139. H. Canadense, L.

Marshes; rare, in Summit Marsh. Aug.

140. H. Canadense, L., var. major, Gray.

Wet places, and marshes; rare. Aug.

Goodwin's Point, where the leaves are sometimes 1–1¼ cm. broad. Summit Marsh, in the water, when the leaves are narrow and red. Rarely on the Montezuma Marshes.

61. Elodea, Juss. Marsh St. Johnswort.

141. E. campanulata, Pursh. (*Elodes Virginica*, Nutt., Man., p. 86.) (H. and C.)

In sphagnum and other marshes; frequent. Aug.

MALVACEÆ.

62. Malva, Linn.

142. *M. rotundifolia, L. Mallow. (H. and C.)

Waste places and roadsides. June–Oct.

143. M. sylvestris, L.

An occasional wayside scape.

Near Ithaca, (*Herb. of C. Humphrey.*) Near White Church. E. of Locke Pond.

144. M. moschata, L. Musk-mallow.

Roadsides and fields; becoming frequent. July–Sept.

Cascadilla Cr. near Cascade Pond. Frequently met with in Enfield and Newfield. Dryden. Frequent in Groton valley.

63. Abutilon, Tourn.

145. A. Avicennæ, Gærtn. Velvet-leaf. (H. and C.)

Waste places; not common. Aug–Oct.

Renwick Farm. Six Mile Cr. near Ithaca, and near old buildings.

64. Hibiscus, Linn.

146. **H. Moscheutos, L.** ROSE-MALLOW. SWAMP HIBISCUS.

Marshes near foot of Cayuga L. Aug.

Union Springs, near Farley's Point; also marsh north of depot. Abundant in a marsh of several acres south of Hill's Branch. A low marshy island of six or eight acres in the middle of Black Lake, has been for several years nearly covered with the noble flowers of this plant. Occasional throughout the Cayuga and Montezuma Marshes.

147. **H. Trionum, L.** FLOWER-OF-AN-HOUR.

In cult. grounds; scarce. **Summer.**

The gardens and orchards on Farley's Pt. Univ. Campus, near the Gymnasium, 1884, (*Miss Hale.*)

ALTHÆA ROSEA, L., the common HOLLYHOCK, has appeared several times by the old roads near the steamboat landing, Ithaca.

TILIACEÆ.

65. Tilia, Linn.

148. ***T. Americana, L.** BASSWOOD.   (**H. and C.**)

Along the ravines, and in rich woods; common. July 12–30.

LINACEÆ.

66. Linum, Linn.

149. **L. Virginianum, L.**   (**H. and C.**)

Dry woods, slopes of ravines; infrequent. July–Aug.

Fall Cr. Coy Glen. Near White Church. Valley near W. Danby.

L. USITATISSIMUM, L. COMMON FLAX.   (**H. and C.**)

Fields and roadsides; probably not permanent. July.

Near Cascadilla Bridge, and W. of Case. Place. Taft Hill, Caroline, and other places.

GERANIACEÆ.

67. Geranium, Linn.

150. ***G. maculatum, L.** CRANES-BILL.   (**H. and C.**)

Ravines, woods and hedges; common. May–June.

151. **G. Carolinianum, L.**

Scarce near Negundo Woods; on South Hill, Ithaca; and by R.R. north of Levanna.

152. **G. pusillum, L.** LITTLE CRANES-BILL.   (**H. and C.**)

Rare, and of recent introduction. Albany St., Ithaca, 1882, and on the Fiske-McGraw grounds, 1885.

153. ***G. Robertianum, L.** HERB ROBERT.   (**H. and C.**)

In moist shaded places, ravines and elsewhere; common. May–Sept.

68. Floerkea, Willd.

154. **F. proserpinacoides, Willd.** FALSE-MERMAID.

In rich woods; scarce. Apr., May.

Negundo Woods. Along Buttermilk Cr., above the upper reservoir. Ellis Hollow. Near Beaver Cr., McLean. Woods N. E. of Freeville. "Near Auburn, N. Y., in wet part of woods where my carex (*C. Careyana*) grows." (*John Carey in Sartwell's Herb.*)

69. IMPATIENS, Linn.

155. *I. pallida, Nutt. PALE TOUCH-ME-NOT. (H. and C.)

In rich woods, along rivulets and ravines; common. July–Aug.

Especially abundant in the vicinity of Woodwardia Swamp and Mud Cr., where a form having flowers of a pale pink with pink spots is found. A white-flowered form was found east of Locke Pond.

156. *I. fulva, Nutt. SPOTTED TOUCH-ME-NOT. JEWEL-WEED.
(H. and C.)

In situations similar to the preceding, but more abundant.

70. OXALIS, Linn.

157. *O. acetosella, L. PINK OXALIS. (C. by Dr. Wright.)

In moss of deep hemlock swamps, and in ravines; frequent. June–July.

Ellis Hollow Sw. Enfield ravine and elsewhere.

158. O. corniculata, L., var. stricta, Sav. [*O. stricta, L., Gray's Man.*, p. 109.] YELLOW WOOD-SORREL. (H. and C.)

Fields, dry woods and elsewhere.

RUTACEÆ.

71. XANTHOXYLUM, Colden.

159. X. Americanum, Mill. PRICKLY ASH. (H. and C.)

Rich soil; not common. May.

Negundo Woods. Renwick Farm slope. Occasional throughout the Negundo valley and along the low points on the lake-shore. Also in Lansing and Dryden.

72. AILANTHUS, Desf.

160. A. GLANDULOSUS, Desf. AILANTHUS.

Introduced abundantly about Ithaca. North of Glen Pond. Cascadilla Cr., below the bridge. E. Hill, near Buffalo St., and University Street. South Hill. The Circus Common. Cayuga Lake, Near McKinney's.

ILICINEÆ. [AQUIFOLIACEÆ, Gray's Man., p. 305.]

73. ILEX, Linn.

161. I. verticillata, Gray. WINTER-BERRY. BLACK ALDER. (H. and C.)

Along streams and marshes; frequent. July.

Its bright red berries especially conspicuous in Ind. Spring Marsh near the lake. Specimens with slender, light colored twigs, are occasionally seen which seem to approach *I. lævigata*, Gray.

74. NEMOPANTHES, Raf.

162. N. Canadensis, DC. "MOUNTAIN HOLLY." (H. and C.)

Borders of sphagnum swamps, where it is frequent. May 15-30.

Larch Meadow, formerly. Enfield Spruce Swamp. The Brookton springs. In all the peat-bogs of Dryden.

CELASTRACEÆ.

75. CELASTRUS, Linn.

163. *C. scandens, L. BITTER-SWEET.   (H. and C.)
Banks of streams, fences, etc.; frequent. June 15-July 15.
Casc. ravine. Along brooks north of Fall Cr., and especially fine fruiting specimens along the declivities of the lake-shore.

EUONYMUS, Tourn.

[E. atropurpureus, Jacq., was discovered by *Fred V. Coville*, in 1885, near Watkins, N. Y., on the rocks by the road on the east side of the valley. It is also in Sartwell's Cat. of 1843.]

RHAMNACEÆ.

76. RHAMNUS, Tourn.

164. R. CATHARTICA, L. BUCKTHORN.
An occasional scape from hedges. May 15-June 15.
Wild south of Fall Cr., W. of Cayuga Southern R. R. On the cliffs south of Levanna, and in Paine's Cr. Planted about the Fiske-McGraw grounds, near Willow Ave., and elsewhere.

165. R. alnifolius, L'Her.   (H. and C.)
In cold springy marshes, where it is frequent. May 15-30.
Indian Spring Marsh. Larch Meadow. Fleming Meadow, and in similar situations within the limits of the Flora.

77. CEANOTHUS, Linn.

166. * C. Americanus, L. JERSEY TEA.   (H. and C.)
Dry woods, along ravines and elsewhere ; common. July.

VITACEÆ.

78. VITIS, Tourn.

167. V. LABRUSCA, L.
Rare, and here probably introduced ; by R. R. near the Glass Works, Ithaca ; in Enfield ravine, in "Lucifer's Kitchen." Roadside one mile south of Mecklenburg.

168. V. æstivalis, Michx. SWEET FROST-GRAPE. (C. by Dr. Wright.)
Dry woods and banks ; frequent. June 20-July 10.
Cascadilla Woods, etc., but especially abundant along the lake shore, where it runs into several forms, some with large clusters, the berries less than 1 cm. in diameter, some with berries 1½ cm. in diam., in small clusters.

169. V. riparia, Michx.   (*V. cordifolia* of Man., p. 112, in part.)
FROST GRAPE.   (H. and C.)
River banks and rich soil, usually ; common. June.
Casc. Cr., and especially abund. about Indian Spring, Negundo Woods and up the Negunena valley ; also along Cayuga Lake shore. This is a very conspicuous vine along the rich alluvial banks of our streams, and often climbs to the tops of the tallest trees. It is dis-

tinguished from the true *V. cordifolia, Michx.*, which scarcely extends to our latitude from the South and West, by the broad, rounded leaf-sinus, the narrow diaphragm separating the pith at the nodes of the branches, (about ¼–½ mm. in thickness), the larger thinner stipules (4–6 mm. long) while the seeds have a less distinct raphe, and the flowering time is from 2 to 4 weeks earlier in the same locality.

[V. cordifolia, Michx., is not certainly known within our limits, although specimens with narrow leaf-sinus have been noticed on Lockwood's Flats and elsewhere. They are all wanting, however, in decisive characters connected with the diaphragms. It is in *Sartw. Herb.*, at least so-named.]

79. AMPELOPSIS, Michx.

170. **A. quinquefolia**, Michx. VIRGINIA CREEPER. (H. and C.)
Fences and in rich soil ; common. July.
This vine covers the cliffs on the lake-shore, below McKinney's, where it is especially beautiful in autumn.

SAPINDACEÆ.

80. STAPHYLEA, Linn.

171. * **S. trifolia**, L. BLADDER-NUT. (H. and C.)
Rich soils along streams and in rocky places ; scarce. May 15–June 15.
Creek banks near Negundo Woods, and in a few localities between this and Ithaca. S. W. corner of Cayuga L. and in a few of the small ravines along the Ledyard, Genoa and Lansing shores. Frontenac Id.

81. ÆSCULUS, Linn.

172. * Æ. HIPPOCASTANUM, L. HORSE-CHESTNUT.
Escaped from cultivation, in Cascadilla ravine below the bridge. On South Hill, and on Rocky Point, Cayuga L. May 15–30.

82. ACER, Tourn.

173. **A. Pennsylvanicum**, L. STRIPED MAPLE. (H. and C.)
Ravines and cold woods on the high hills ; frequent. May 20–June 20.

174. **A. spicatum**, Lam. MOUNTAIN MAPLE. (H. and C.)
Ravines, woods and rocky hillsides ; frequent. May 20–June 20.
Specimen south side of Six Mile Cr. Narrows nearly 10 cm. in diameter.

175. * **A. saccharinum**, Wang. SUGAR MAPLE. (H. and C.)
Ravines and woods ; common. May 12–30.
Large groves of these toward Cortland ; also a large element in the woods of Dryden, Caroline, Danby, Enfield and Newfield.

176. **A. saccharinum**, L., var. **nigrum**, Gray. BLACK MAPLE. (Seeming like a distinct species.)　　　　　(C. by Dr. Dwight.)
Richer soil than the preceding ; not common.
Toward Enfield (*Jordan*)! Negundo Woods, Indian Spring, and at the Corner-of-the-lake, Taughannock, Paine's Cr., and Big Gully ravines and elsewhere, chiefly near the level of the lake.

177. **A. dasycarpum**, Erhart. WHITE MAPLE. SILVER MAPLE.
(**C.** by Dr. Wright.)
Mostly in alluvial soil on the lower levels; frequent. Mar.–April.
Marshes near the Lake (*Jordan*)! Negundo Woods and south.
Occasional in the swampy regions of Lansing and Dryden, and very
abund. in the woods on the Cayuga and Montezuma Marshes. One
of the earliest bloomers.

178. * **A. rubrum**, L. RED MAPLE. SOFT MAPLE. (**H.** and **C.**)
Chiefly on the hillsides or in upland woods and swamps; common. April.

83. NEGUNDO, Mœnch.

179. **N. aceroides**, Mœnch. BOX-ELDER.
In "Negundo Woods" at a bend of the creek, 1½ miles south of
Ithaca. Discovered here by Professor Branner, when a student,
June, 1873. The only station reported for N. York State until 1885,
when Negundo was reported from near Binghamton by Dr. Mill-
spaugh in the Bull. of the Torr. Club. There were, when first discovered, about 20 trees, at the Ithaca station, ranging from ⅓ to ⅓
meter in diam., and without doubt indigenous. Since then most of
them, together with the magnificent old Buttonwoods and Elms,
have been cut down. There are some young trees, both here, on
the banks of the creek below the railroad bridges, and near the
McGraw barns.

ANACARDIACEÆ.

84. RHUS, Linn.

180. **R. typhina**, L. STAGSHORN. (**H.** and **C.**)
Ravines and hillsides; common. July 1–15.

181. * **R. glabra**, L. SMOOTH-SUMAC. (**H.** and **C.**)
Banks and hillsides; frequent. July 10–30.
Slopes of East Hill, South Hill, and abundant on shore of Cayuga
Lake.

182. **R. venenata**, DC. POISON SUMAC OR P. DOGWOOD. (**H.** and **C.**)
Marshes; not common. July 1–20.
Indian Spring Marsh, and occasional in the sphagnum swamps.

183. * **R. Toxicodendron**, L. POISON IVY. (**H.** and **C.**)
Low grounds and rocky places; common. June–July.

184. **R. aromatica**, Ait. AROMATIC SUMAC. (**H.** and **C.**)
Rocky declivities, where it is frequent. May.
Fall Cr., near Triphammer Falls, and elsewhere. Found on
South Hill, in Six Mile Cr., in all the ravines, on the high "Pinna-
cles" at West Danby and White Church, and especially abund. on
the E. shore of the lake.

LEGUMINOSÆ.

85. BAPTISIA, Vent.

185. **B. tinctoria**, R. Br. WILD INDIGO.
In Junius, Seneca Co. (*Sartwell in Herb. and C.*)

86. LUPINUS, Tourn.

186. **L. perennis, L. LUPINE.**
Dry woods, where it is not uncommon. May 10–June 10.
Cascadilla woods, near the Gymnasium, and the Psi-U. House.
Near all our ravines, also in woods on the high hills of Newfield,
Danby and Caroline. "Junius." (*Sartw. in H. and C.*)

87. MEDICAGO, Linn.

187. **M. SATIVA, L. LUCERNE.**
Sparingly introduced on the Campus, also on the Pine Knoll in
Case. Woods, 1871, from which it has recently disappeared. On
Cayuga L. shore at Canoga.

188. **M. LUPULINA, L. BLACK MEDICK.**
A weed. Chiefly near the large towns, the lake and R. Roads.
May–Sept. Junius. (*Sartw. Herb. and C.*)

88. MELIOTUS, Tourn.

189. *M. OFFICINALIS, Willd. YELLOW MELILOT. (H. and C.)
Near the shore of Cayuga Lake. June–Aug.
Ledyard, (*Herb. Thomas.*) Rare near the Inlet. McKinney's.
Abund. from Union Springs to Montezuma.

190. *M. ALBA. Lam. SWEET CLOVER. (H. and C.)
Very common along Cayuga L. and the gravelly banks of the
creeks. Ledyard. (*Herb. Thomas.*) June–Oct.

89. TRIFOLIUM, Linn.

191. *T. PRATENSE, L. RED CLOVER. (H. and C.)
Common. June–Oct.

192. T. HYBRIDUM, L. ALSIKE. (Intermediate in aspect between
T. pratense and *T. repens.*
Roadsides; fields; frequent. June–Aug.
First noticed as a scape, at McLean and Ithaca, 1878. Frequent-
ly cultivated in Danby, Enfield, Ovid and elsewhere.

193. *T. REPENS, L. WHITE C. (H. and C.)
Common. June–Oct.

194. T. AGRARIUM, L. YELLOW OR HOP-CLOVER. (H. and C.)
Gravelly soil; chiefly on the hills. June–July.
Scarce; near the Valley Cemetery, on the hills near White Church'
and West Danby; but abundant on the hill-farms in eastern part of
Danby.

195. T. PROCUMBENS, L. LOW HOP CLOVER.
Of recent introduction; Univ. Campus, 1882; in bloom in July,
in pasture north of the Glen Pond; also in Ithaca Cemetery.

LOTUS, Linn.

L. CORNICULATUS, L.
Appeared in the turf on the Fiske-McGraw grounds, in bloom
Aug. 1885.

90. TEPHROSIA, Pers.

196. **T. Virginiana, Pers. HOARY PEA.**
Dry sandy soil. "Junius," (*Sartwell, H. and C.*)

91. ROBINIA, Linn.

197. * R. PSEUDACACIA, L. LOCUST-TREE.
Ravines and hillsides ; frequent. June 1-15.
Cascadilla, and most of the larger ravines, and abundant along the smaller ravines bordering on the lake.

198. R. VISCOSA, Vent. CLAMMY LOCUST.
Escaped ; at Aurora, a few shrubs near the R. R. track. June 15-30.

92. ASTRAGALUS, Linn.

199. **A. Canadensis**, L. MILK VETCH. (**H**. and **C**.)
Cayuga Lake shore ; infrequent. July.
Corner-of-the-lake, Bushy Point, Crobar Point, and occasional on both shores as far as Cayuga Bridge.

200. **A. Cooperi**, Gray. (**H**. and **C**.)
Cayuga L., on the rocky banks ; rare. July.
South of Shurgur's Glen, (*F. E. Hinc*)! North of Ludlowville. A considerable number on the cliff south of Levanna.

93. DESMODIUM, DC. TICK TREFOIL.

201. **D. nudiflorum**, DC. (**H**. and **C**.)
Dry woods ; frequent. July 20, Aug.
A white-flowered form on Thacher's Pinnacle, W. Danby. A small form (approaching *D. pauciflorum, DC.,*) with leaves scattered along the slender ascending stem, is not uncommon, especially on the higher hills.

202. **D. acuminatum**, DC. (**H**. and **C**.)
In richer woods than the preceding ; common. July 15-Aug.

203. **D. rotundifolium**, DC.
Sunny copses ; frequent. Aug.
Casc. Woods, Fall Cr. and other ravines. White Church. Thacher's Pinnacle. "Junius," (*Sartwell H. and C.*)!

[**D. canescens**, DC.
"Seneca L. *Beck in Herb.*" *Paine's Cat.* p. 73. Gorham. *Sartwell in H. and C.* "Abund. in gully E. of Penn Yan." (*Dr. Wright.*)]

204. **D. cuspidatum**, Torr. and Gray. (**H**. and **C**.)
Rich dry woods ; not common. Aug.
Casc. Woods. Coy Glen. White Church, and elsewhere.

205. **D. Dillenii**, Darlingt. (**C**.)
Along ravines and in dry woods ; frequent. Aug.
Varies considerably, the more downy forms approaching *D. viridiflorum*, Beck. Paine gives "Seneca L." as a station for *D. viridiflorum*, Beck, on authority of Sartwell, but I have not yet seen the true form in this region.

206. **D. paniculatum**, DC. (**H**. and **C**.)
In woods and ravines ; frequent. Aug. 10-30

207. **D. Canadense**, DC. (**H**. and **C**.)
Banks of creeks, and on the lake shore ; abundant. Aug, Sept.

208. **D. rigidum,** DC. (C. by Dr. Wright.)

Open thickets and hillsides ; scarce. Aug.

Near ravine north of Enfield ravine. Near White Church. A form (possibly a hybrid *D. rigidum DC. × D. Marilandicum, Boott,*) with smoothish stem and leaves 3–4 cm. long, nearly smooth, occurs at White Church.

209. **D. Marilandicum,** Boott. (H. and C.)

Open thickets ; not common. Aug.

Coy Glen (*F. B. Hinc.*) Near White Church. var. *β.* Torr. and Gray, with larger leaflets occurs near Fall Creek mill-pond, and near White Church.

[D. ciliare, DC. At Penn Yan (*Sartwell, H. and C.*)]

94. LESPEDEZA, Michx. BUSH CLOVER.

210. **L. repens,** Bart. (*L. repens,* Torr. and Gray and *L. procumbens,* Michx. See Gray's Man., p. 137. (H. and C.)

Dry banks ; rare. Aug.

Banks of ravine north of Enfield ravine. (S. 4½ × W. 6½.) At same place, is a form approaching *L. violacea,* Pers.,in the character of loments.

211. **L. violacea,** Pers. (*L. violacea* var. *divergens,* of Man., p. 137) (H. and C.)

The common form bears numerous thin leaves and rather glomerate clusters of purple flowers, (but is quite different from *L. reticulata,* Pers.,) Frequent in dry woods. Aug.–Sept.

212. An open loosely panicled form with handsome flowers in pairs, large, nearly 1 cm. long, and with large ovate acuminate pods ; occurs on Farley's Point.

213. A form with thin leaves and slender stem, but with petioles and stem pubescent with spreading hairs, the flowers pink or only the vexillum purplish, and the short racemes peduncled, seems to belong here although it may be a hybrid between this species and the next. It occurs in Casc. Woods and on banks of ravine north of Enfield ravine.

214. **L. Stuvei,** Nutt.

Dry banks ; scarce. Aug.–Sept.

Hillside near White Church. Banks of ravine north of Enfield ravine.

215. **L. hirta,** Ell. (H. and C.)

Dry banks and woods ; common. Aug.–Sept.

216. **L. capitata,** Michx.

Rare, on the wild bank beyond the Fleming S. H. Aug. 20–Sept. 20.

(Milo, roadside. *Dr. Wright.*)

95. VICIA, Tourn.

217. **V. SATIVA,** L. COMMON VETCH. (H.)

Roadsides and banks ; occasional. June–Sept.

In the valley south of Ithaca, along Cayuga L. and elsewhere.

218. **V. Cracca**, L. (H. and C.)

On the sand bank west of Mr. Howard Williams', since 1871 (abundant by R. R. near Cortland depot.) July; appearing as if introduced here.

219. **V. Caroliniana**, Walt. (C.)

Ravines and banks; common. May 10–30.

220. **V. Americana**, Muhl. (H. and C.)

Only along Cayuga L. and its marshes, where it is frequent. June–July.

96. LATHYRUS, Linn.

221. **L. venosus**, Muhl. "? Geneva" in Sartwell's Herb." "Cayuga L.," (*Herb. Thomas.*)

222. **L. ochroleucus**, Hook. (H. and C.)

Hillsides and thickets; not common. May.

Fall Cr., and other ravines, also ravines and woods of Cayuga lake shore.

223. **L. palustris**, L. MARSH VETCHLING. (H. and C.)

Shores and borders of marshes; not common. July.

Near the Corner-of-the-lake. Shore of Cayuga L., at Marion's, on Bushy Point, and occasional to Cayuga Marshes.

224. **L. palustris**, L. var. **myrtifolius**, Gray. (H. and C.)

Shore of Cayuga L.; not common. July–Aug.

Near Marion's. Near Treman's. Myers Point. Near Union Springs.

97. AMPHICARPÆA, Ell.

225. *****A. monoica**, Nutt. HOG PEA-NUT. (H. and C.)

Thickets, shores, etc.; common. Aug.–Sept.

98. APIOS, Boerheave.

226. **A. tuberosa**, Mœnch. GROUND-NUT. (H. and C.)

Rich alluvial soil; abundant in the Negnæna valley, and along the lake. Aug.

Cascadilla ravine. Negundo Woods and south. Myers Point. Cayuga Marshes.

99. CASSIA, Linn.

227. **C. Marilandica**, L. WILD SENNA.

Rare. Roadside near the fork of the Slaterville and Brookton roads, E. of Ithaca.

100. GYMNOCLADUS, Lam.

228. **G. Canadensis**, Lam. KENTUCKY COFFEE-TREE. (H. and C.)

Rare; along the borders of Cayuga L. May 25–June 10.

"At the bottom of a ravine near the borders of Cayuga L. (*Dr. Alex. Thompson*)," (see Torreys Flora, I, p. 191); also in Dr. Thompson's Cat. of 1841. It is quite likely the precise locality was Lockwoods Flats at the mouth of Big Gully, where Professor Thomas of Union Springs remembers to have seen it many years ago. "Ithaca (*in Herb. Van Duzer*)," (see Paine's Cat. p. 75). Town of Lansing "near Cayuga L." at McKinney's, (*Hon. H. B. Lord in*

19th Report of Regents, 1865, p. 77), who says there are one large (18 in.) and two small trees, which the people near suppose to be *mahogany* trees. The large and one small tree are still standing (1885). Three middle-size trees near the brook at Lake Ridge Sta. ! Three rather large trees apparently indigenous near the Lehigh Valley R. R. round-house at Ithaca!

("On Seneca L. at Cachong Cr.—one tree 18 in. diam." (*J. Hall*, in Torr. Flora.) "Seneca Lake," (*Sartwell Herb.*) Near Fayette, Onondaga Co., (*Professor Underwood, letter of Feb.* 1886), "Onondaga Co." (*Mrs. Rust, Bull. of Torry Bot. Club, Vol. IX.*)

The above are the only stations known to me in N. Y., where *Gymnocladus* is presumably indigenous.

101. GLEDITSCHIA, Linn.

229. * G. TRIACANTHUS, L. HONEY-LOCUST. (H and C.)
Hillsides and banks of streams ; occasional. June 20-30.
Near Casc. Cr., on the slope of East Hill, Six Mile Cr., South Hill and elsewhere near Ithaca.

G. MONOSPERMA, Walt.
A single young tree growing spontaneously, by one of the roads south of the steamboat landing.

ROSACEÆ.

102. PRUNUS, Tourn.

230. P. PERSICA, L. PEACH.
Occasional, by roadsides. Near Indian Spring, toward Buttermilk Falls, and elsewhere. Probably it rarely self-seeds. First week in May.

231. P. Americana, Marshall. WILD PLUM. (H. and C.)
Ravines and thickets ; common. May 1-20.
Two forms occur ; one with gray, slender branches, larger flowers, and glandular calyx-lobes, occurs in ravines, often on swamp-borders. The other, with dark, stouter twigs, smaller flowers than the preceding, the calyx-lobes silky within, slightly or not at all glandular, occurs in hedges or more exposed places than the other.

232. P. DOMESTICA, L. GARDEN PLUM.
Escaped ; not uncommon. May 15—25.
Cornell's Woods, Renwick-Farm slope, South Hill and elsewhere.

233. P. SPINOSA, L. SLOE.
Apparently rare. Shurgur's Point, Cayaga Lake. May.

234. P. AVIUM, L. ENGLISH CHERRY.
Ravines, and banks of Cayuga Lake ; frequent. May 15-30.
All ravines near Ithaca, and especially abundant on both shores of the Lake, where it forms small thickets and groves on the cleared banks. Not uncommon on Turkey Hill, also in Enfield, Newfield and Dryden.

235. * P. CERASUS, L. SOUR CHERRY.
Escaped to roadsides, usually not far away from farm-houses ; infrequent. May 15-30.

Dryden Road, near Heustis St., Renwick Farm, Danby and else-where.

236. **P. pumila**, L. DWARF CHERRY.
Rare. Only known on the hummocks of South Hill Marsh,! where it was detected by Mr. Curtice, in 1882. May 25–June 10.
The habitat is unusual, and remarkable in view of the fact that it has been vainly sought for on the lake-shore, where it would be expected. Its drupes at our station are usually destroyed by *Exoascus Pruni, Fck'l*, a fungus.

237. **P. Pennsylvanica**, L. WILD RED CHERRY. (C.)
On newly cleared land and in ravines ; common. May 15–20.

238. * **P. Virginiana**, L. CHOKE CHERRY. (H. and C.)
Ravines and rocky banks ; common. May 10–25.

239. * **P. serotina**, Ehrh. WILD BLACK CHERRY. (H. and C.)
Woods ; frequent. May 20 June 30.

103. SPIRÆA, Linn.

240. **S. salicifolia**, L. MEADOW SWEET. (H. and C.)
Marshes and low grounds ; common. Aug.–Sept.
S. LOBATA, Murr.
A bed of this plant has been growing vigorously for some years on the marsh north of Ithaca. July.

104. NEILLIA, Don.

241. **N. opulifolia**, Benth. and Hook. (*Spiræa opulifolia, L.*) NINE-BARK.
Cayuga L. shore, (*Jordan*), where it is abundant ! June 10–30.
"Ithaca," (*Sartwell Herb.*) Ledyard, (*Herb. J. J. Thomas.*)
It is especially abundant and beautiful on the shales of the High Cliffs, north of Kings Ferry. It occurs in Fall Cr., below the Mir-ror Falls, and in Taughannock ravine.

105. GILLENIA, Mœnch.

242. * **G. trifoliata**, Mœnch. BOWMAN'S ROOT. (H. and C.)
Dry woods, with *Ceanothus*, etc. ; frequent. June.
Casc. Woods, on the knoll near the Armory, and in similar situa-tions.

106. DALIBARDA, Linn.

243. **D. repens**, L. (*Rubus Dalibarda, L.*) (C. by Dr. Wright.)
Borders of sphagnum swamps and hemlock woods. July–Aug.
Enfield ravine. Near Michigan Hollow Swamp, and not uncom-mon near the low woods and swamps, Freeville and McLean. Led-yard, (*Herb. J. J. Thomas.*)

107. RUBUS, Tourn.

244. * **R. odoratus**, L. FLOWERING RASPBERRY. (H. and C.)
Rocky or gravelly banks ; common. June–Aug.
Specimens with white and with pink flowers on west shore of the lake. (*F. B. Hine*, 1876. *F. T. Wilson*, 1882)!

245. **R. triflorus**, Rich. (C.)
Moist woods and hedges; frequent. May.

246. **R. neglectus, Pk.** (See 22d Rep. of Regents, p. 53.) CREAM-BERRY.
Hillsides and pastures ; not uncommon. June–July.
Cayuga L., west shore near Marion's. W. Danby. Near Summit Marsh. Casc. Cr., above Judd's Falls.
It is distinguished by its stems recurved, (like *R. occidentallis*), and armed with unequal straight prickles, its pedicels armed with numerous slender prickles intermingled with stiff glandular hairs, its hispid calyx, its fruit dark red, intermediate in color between *R. strigosus* and *R. occidentalis*. It is distinct, but seems like a hybrid.

247. **R. strigosus, Michx.** RED RASPBERRY. (H. and C.)
Light soil, especially on newly cleared lands ; common. June.

248. **R. occidentalis, L.** BLACK RASPBERRY. (H. and C.)
Hillsides, fences, etc.; common. June.

249. **R. villosus, Ait.** BLACKBERRY. (H. and C.)
Woods and thickets ; common. May–June.

250. **R. villosus, Ait. var. humifusus, Gray.** (H.)
Cascadilla Cr., Taughannock and elsewhere ; not common.

251. **R. villosus, Ait. var. frondosus, Gray.**
South Hill.

252. **R. Canadensis, L.** DEWBERRY. (C.)
Rocky places or old fields ; frequent. June.

253. **R. hispidus, L.** SWAMP BLACKBERRY. (H. and C.)
Marshes and borders of sphagnum swamps ; frequent. June.
South Hill Marsh, Larch Meadow and other cold, wet marshes.

108. GEUM, Linn.

254. **G. album, Gmelin.**
Hedges and moist banks ; frequent. July.

255. **G. Virginianum, L.** (H. and C.)
With the preceding ; frequent. June, July.

[G. macrophyllum, Willd.
S. H. Wright finds this in a gully E. of Penn Yan.]

256. **G. strictum, Ait.** "BLACKBURR," (local name.) (H. and C.)
Fields and hedges ; common. June 25–July 15.

257. ***G. rivale, L.** PURPLE AVENS. (H. and C.)
Marshes ; frequent. May 20–June.
South of Fall Creek, near Forest Home. Ellis Hollow swamp. Throughout marshes of Dryden and Danby.

109. WALDSTEINIA, Willd.

258. ***W. fragarioides, Willd.** BARREN STRAWBERRY. (H. and C.)
Ravine banks, thickets and woods; common. May 1–20.

110. FRAGARIA, Tourn.

259. **F. Virginiana, Ehrh.** WILD STRAWBERRY. (H. and C.)
Fields and pastures ; common. May.

260. **F. Virginiana.** Ehrh.   var. **Illinoiensis,** Gray.
    In richer soil; less common. May.
    Six Mile Cr. South Hill and near Cayuga L.

261. **F. vesca,** L. WOOD S.   (**H.** and **C.**)
    In woods and swamps or recently cleared land; frequent. May–
    June.

262. The var. **alba,** the "*Indian Strawberry,*" is at Sherwood, and is
    not uncommon in that region, according to Mrs. Professor Brun.
    Its fruit is white and its leaves thicker and more shining than in
    the type. Also near Beebe Pond, (*Professor J. H. Comstock.*)

III. POTENTILLA, Linn.

263. P. RECTA, L.
    Fields; scarce. July.
    Near Union Springs, 1875!

264. **P. Norvegica,** L.   (**H.** and **C.**)
    Fields; common. July–Oct.

265. *P. Canadensis, L. FIVE-FINGER.   (**H.** and **C.**)
    Fields; common. May–July.

266. **P. Canadensis,** L. var. **simplex,** Torr. and Gr.
    Dry banks; common.

267. **P. argentea,** L. SILVERY CINQUE-FOIL.   (**C.**)
    Gravelly or rocky hillsides. Not common. June–July.
    Campus, south of Morrill Hall. South Hill. Lansing. Near Lud-
    lowville. The cedar knoll near High Cliffs. E. of McLean, and
    elsewhere.

268. **P. arguta,** Pursh.   (**C.** by Dr. Wright.)
    Rocky or dry banks; rare. May 20–June 15.
    South of Burdick's Glen, near the shore. Beyond the Fleming
    S. H., with *Lespedeza capitata,* and found in 1881 introduced in a
    grass-field south of McLean. (Watkins, N. Y., E. side of valley.)

269. *P. Anserina, L. SILVER-WEED.   (**H.** and **C.**)
    Sandy shores, where it is frequent. June–July.
    Abundant at the head of Cayuga Lake (*Jordan*)! and on all the
    sandy points to Cayuga and Montezuma Marshes. Summit Marsh.
    Cortland Marl Ponds and Cayuta L.

270. **P. fruticosa,** L. HARD-HACK, OR SHRUBBY CINQUE-FOIL.
    Sphagnum swamps; rare. June–Aug.
    Round-Marsh (eastern.) Locke Pond, above bridge. Summit
    Marsh, east side, a few shrubs at each station. Junius, (*Sartwell
    in Herb. and C.*), where it is abundant about Newton's Ponds!
    ("Yates Co., (*Sartwell*)" in Torrey's Flora, I, p. 210.)

271. *P. palustris, Scop. MARSH FIVE-FINGER.
    Marshes; infrequent. June 20–July 15.
    Round-Marsh, (middle one.) By the pond S. E. of Chicago
    Sta. Locke Pond, (E. side.) Cayuga Marshes, from Cayuga to
    Montezuma. Cayuta L. (N. W. shore.) Junius, (*Sartwell H. and C.*)

112. AGRIMONIA, Tourn.

272. **A. Eupatoria**, L. AGRIMONY.   (H. and C.)
 Fields ; common. July–Sept.

273. **A. parviflora**, Ait.
 Marshy places ; rare. Aug.–Sept.
 W. Danby ; collected by *F. C. Curtice*, June, 1882. Near Freeville, (one mile west), July and Sept. 1882. (Previously reported
 from N. Y., only from Rockland Co., in the 20th Rep. of Regents,
 p. 403.)

113. POTERIUM, Linn.

274. **P. Canadense**, Gray. CANADIAN BURNET. (C. by Dr. Wright.)
 Sedgy sphagnum swamps ; rare. Aug.
 Larch Meadow, 1873! Along Mud Creek, Freeville. Locke
 Pond, in marsh.

114. SANGUISORBA, Linn.

[S. OFFICINALIS, L. Occurs well established along the West Shore
R. R. east of Clyde Sta.]

115. ROSA, Tourn.

275. **R. blanda**, Ait. EARLY WILD-ROSE.   (H.)
 On the rocky banks of Cayuga L., where it is abundant. June.
 Occasional up the Neguæna valley. Forms with pubescent leaves
 approaching the western var. **pubescens**, Crepin, occur near the
 lake shore.

276. **R. Carolina**, L. SWAMP ROSE.   (H. and C.)
 Swamps and low grounds ; common. July.

277. **R. lucida**, Ehrh. SHINING ROSE.
 North of Eddy Pond, and probably not rare. Near Junius Ponds.
 "Newfoundland to E. New York, and Penn.," (*Sereno Watson in
 "Revision of Roses of North Amer.*," Proc. of Amer. Acad., Vol.
 XX, 1885, p. 147.) Mr. Watson writes that the above specimens
 are well marked, and from the most western station known to him.
 This species is from $\frac{1}{3}-1\frac{1}{2}$ m. tall, its leaflets more coarsely toothed
 than in *R. Carolina*, dark green and thickish, its stipules more or
 less dilated, and its spines stout and straight, or when old usually
 hooked.

278. **R. humilis**, Marshall. (*R. lucida*, of Man., p. 158, for most part.)
 DWARF ROSE.   (H. and C.)
 Dry woods and banks ; common. June–July.
 A taller, nearly unarmed form of this species grows on the lake
 shore banks south of McKinney's Sta., and elsewhere.

279. **R. setigera**, Michx. PRAIRIE ROSE.
 Perhaps introduced in this region from the West ; scarce. July.
 Cascadilla ravine below the bridge ; also north of Glen Pond. Six
 Mile Creek,—"amphitheatre." South Hill by R. R., near the
 Reservoir. Field near Utts Point,—always in wild places.

280. **R. CINNAMOMEA**, L. CINNAMON ROSE. Persists frequently by
 roadsides near houses. Seemingly spontaneous by road north of
 Cayuta L.

281. * R. RUBIGINOSA, L.   SWEET-BRIER.   (H. and C.)
Pastures and hillsides; frequent. July.

       116. PIRUS, Linn.

282. P. MALUS, L.   APPLE.
Self-seeding and frequent. May 1–25.
Casc. Cr., near Eddy Pond. Slopes of E. Hill and South Hill.
Lake-shore and elsewhere.

283. P. COMMUNIS, L.   PEAR.
Self-seeding occasionally. May 1–20.
North of Eddy Pond. Six Mile Cr., near Ferris Brook and else-
where.

284. P. coronaria, L.   WILD CRAB-APPLE.   (H. and C.)
Hillslopes, frequently in moist soil; not uncommon. May 15–
June 1.
The University Grove on the President White grounds (indigenous.) E. Hill slope. Renwick-Farm slope, and north of Fall Cr.
Six Mile Creek, (a tree near the Valentine place, about 25 cm. in
diam.), rather abundant on South Hill, near the "Incline," and
beyond S. S. Sta. 420. On West Hill. In Danby, Newfield, and
not uncommon near Union Springs. When in bud and partly
in bloom this is the most beautiful and fragrant of our native
Rosaceæ.

285. P. arbutifolia, L., var. erythrocarpa, Gray.   CHOKE-BERRY.
                                              (H. and C.)
South Hill, near the R. R., on the rocky bank of the South
Quarry Cr. Also near South Hill Marsh; not common. June.

286. P. arbutifolia, L., var. melanocarpa, Gr.   CHOKE-BERRY.
Chiefly on the borders of sphagnum swamps; abundant. May 20–
June 15.
Indian Spring Marsh. Larch Meadow, and cold marshes of Dry-
den, Danby, Enfield and elsewhere.

287. P. Americana, DC.   MOUNTAIN ASH.
Indigenous, along the borders of sphagnum swamps; rare. June.
Woodwardia Sw. Malloryville Marsh. Round-Marshes. Michigan Hollow Swamp. Usually a mere shrub in our Flora. Occasionally cultivated, and small specimens—presumably scapes—are
frequently found in upland woods, and by roads.
P. SAMBUCIFOLIA, Schlect. and Cham.; rarely escapes to roadsides.

       117. CRATÆGUS, Linn.

288. C. OXYCANTHA, L.   ENGLISH HAWTHORN.   (H.)
Escaped to pastures; rare. June 1–25.
Cascadilla Cr., north of Cascade Pond, formerly. Grassy levels
in Big Gully ravine. Pasture near Farley's Point. Many fine
plants in cultivation on Farley's Point.

289. * C. coccinea, L.   SCARLET THORN.   (H. and C.)
Thickets and hillsides; common. May 15–30.
Many of our plants have appressed-hairy, instead of "glabrous"
leaves.

290. **C. coccinea**, L. var. **macracantha**. (*C. glandulosa*, Willd. of most early American works. *C. macracantha*, Lodd, probably. Referred by Torr. and Gray, (1840.) to *C. coccinea*, L.; see *Flora*, I, p. 465.)

Hillsides and stony soil ; frequent. May 20–30.

Near the Campus Brook. South Hill, beyond the " Incline," and beyond S. S. 420. Both shores of Cayuga L., from Rocky Point, north. Specimens from Union Springs are intermediate between this form and the true *C. coccinea ;* nevertheless, this form, as seen in Central N. Y. and elsewhere, is so distinct from the type, that to fail to distinguish it by name in some way, gives rise to more or less confusion. It differs from the ordinary thin-leaved *C. coccinea*, L., in the following particulars : Branches usually long, straggling ; branchlets stout, reddish, with globose or oblate, red shining buds. Thorns very long, 8–12 cm., rarely more, Leaves broadly obovate-cuneiform, lobed and doubly-serrate above, tapering by a wedge-shaped base into a stout petiole, coriaceous, smooth, shining, veins prominent. Calyx-lobes narrowly oblong or lanceolate, pectinately-glandular, while the broader calyx-lobes of *C. coccinea* are only sparsely glandular. Flowers are smaller and more abundant than in *C. coccinea*. Fruits smaller, not so fleshy, seeds bony and larger.

291. **C. tomentosa**, L. BLACK THORN. (H. and C.)

Hedges and moist soil ; frequent. June 5–15.

Frequently without thorns, its calyx-lobes glandular-pinnatifid.

292. **C. tomentosa**, L., var. pyrifolia, Gray.

Ledyard, 1827. (*Herb. J. J. Thomas*,) "Near Auburn" (*J. Carey in Flora of N. Y.* I, p. 222.) North and South of Union Springs. Its leaves are rather ample, spatulate, long-petioled, somewhat shining and almost regularly serrate. It is distinct from the type in time of flowering (May 25–June 5), and in the entire glandless calyx-lobes (agreeing in both these points with the following variety.)

293. **C. tomentosa**, L. var. punctata, Gray. (C.)

Hillsides, fields, more rarely in woods ; abundant. May 25–June 5. Fruit usually a dull crimson, but yellow pomes occur near Etna, and Forest Home; in White Church valley and near Cayuta L.

294. **C. Crus-galli**, L. COCK-SPUR THORN. (H. and C.)

Pastures and fences.

Frequent near Cayuga L. from a region south and east of Ludlow-ville to Montezuma ; also west and north of Sheldrake and Canoga ; elsewhere absent.

118. AMELANCHIER, Medic.

295. *A. Canadensis, Torr. and Gray. (including var. *botryapium*.)
JUNE-BERRY. SHAD-BUSH. (H. and C.)

Ravines and woods ; frequent. May 1–20.

On Univ. farm (a large tree, the young leaves hairy.) Fall Cr., Case. Cr., Eddy Pond, Six Mile Cr., Big Gully, etc., (all with young leaves nearly glabrous.)

296. **A**. Canadensis, Torr. and Gray. var. oblongifolia, Gray. (C.)

Ravines and woods ; common. May 1–20.

297. **A. Canadensis**, Torr. and Gray. var. rotundifolia, Gray.

(H. and C.)

Rocky banks and cliffs; not common. May 20-June 10.

Fall Cr. ravine (*F. L. K.*)! Cayuga L.; Kings Ferry, and north of Crobar Pt. Caroline on the cliffs near White Church. W. Danby, on Thacher's pinnacle.

298. **A. Canadensis**, Torr. and Gray; (form.)

(In sphagnum marshes, and agreeing with var. *oligocarpa* in character of the leaf and length of petals, but racemes usually have 4-6 flowers.) Round Marshes. South Hill Marsh. May 10-20

SAXIFRAGACEÆ.

119. SAXIFRAGA, Linn.

299. **S. aizoides**, L. YELLOW MOUNTAIN-SAXIFRAGE.

Dripping cliffs; very rare. June 15-30.

Cliffs of Taughannock ravine, south side, below the falls; detected Oct. 1880. Elsewhere in N. Y. only at Fish Cr., Oneida Co., and at Portage and in Warsaw Glen, 1871, (*Dr. Jordan*)! also from Portage, (*Buffalo Cat.* p. 97.)

300. * **S. Virginiensis**, Michx. EARLY SAXIFRAGE. (H. and C.)

Rocks along ravines and brooks; common. Apr.-May.

301. **S. Pennsylvanica**, L. SWAMP SAXIFRAGE. (C.)

Wet cold swamps; frequent. May.

Indian Spring Marsh, Larch Meadow and elsewhere.

120. TIARELLA, Linn.

302. * **T. cordifolia**, L. FALSE MITRE-WORT. (H. and C.)

Ravines and cool banks; abundant. May-June.

121. MITELLA, Linn.

303. * **M. diphylla**, L. BISHOP'S-CAP. MITRE-WORT. (H. and C.)

Ravines and damp thickets; common. May-June.

304. **M. nuda**, L. (H. and C.)

Borders of sphagnum marshes; infrequent. May 20-June 10. Ellis Hollow Swamp. Malloryville Marsh, (*F. P. Weeks*.)! Freeville, Fir-tree Swamp and Mud Cr. Round Marshes. Michigan Hollow. Bear Sw.

122. CHRYSOSPLENIUM, Tourn.

305. **C. Americanum**, Schw. GOLDEN SAXIFRAGE. (H. and C.)

Cold, springy places; frequent. Apr.-May.

123. PARNASSIA, Tourn.

306. **P. Caroliniana**, Michx. GRASS OF PARNASSUS. (H. and C.)

In springy places in all the ravines and in peat-bogs; frequent. Aug.-Sept.

124. RIBES, Linn.

307. **R. Cynosbati**, L. PRICKLY GOOSEBERRY. (H. and C.)

Dry woods, banks and cliffs; common. May.

Very common along the lake shore cliffs, when occasionally the berries are smooth; and the stems are often pendant, 3-4 m.

308. **R. oxycanthoides,** L. (*R. hirtellum, Mx.,* Man., p. 164)
SWAMP G. (**H.** and **C.**)
 Low grounds ; not uncommon. May 12-30.
 Near Negundo Woods, Larch Meadow, and the marshes of Dryden.

309. **R. rotundifolium,** Michx.
 Rocky slopes ; not common. May 12-25.
 South Hill near the "Incline." Near the "Nook" green-house.
 Road toward Enfield Falls ? (*H. E. Summers.*) Groton.

310. **R. lacustre,** Poir.
 Deep swamps ; scarce. June.
 "Freeville Sw." (*Jordan and Copeland,* 1872,—not in herb.)
 Swamp near Round-Marsh. Michigan Hollow. Ellis Hollow.

311. **R. prostratum,** L.'Her. FETID CURRANT.
 Cold woods and deep swamps ; frequent. May 10-30.
 Dart Woods and Ellis Hollow swamp, and in similar situations.

312. **R. floridum,** L. BLACK CURRANT. (**C.**)
 Moist banks and along creeks ; frequent. May.

313. **R. RUBRUM,** L. GARDEN CURRANT.
 Escaped. Near Indian Spring. South Hill ; scarce.

314. **R. rubrum,** L. var. **subglandulosum,** Maxim. WILD RED-CUR-
RANT.
 Cold woods and deep swamps ; not common. May 10-25.
 Ellis Hollow swamp. Dryden-Lansing swamp. (*Jordan*)! Free-
ville and Beaver Creek. Michigan Hollow. North of Cayuta L.

CRASSULACEÆ.

125. PENTHORUM, Gronov.

315. **P. sedoides,** L. DITCH STONE-CROP. (**H.** and **C.**)
 Low grounds ; common. Aug.-Sept.

126. SEDUM, Tourn.

316. **S. ACRE,** L. GOLDEN OR MOSSY STONE-CROP.
 Escaped sparingly ; North side Fall Cr. below Ithaca Falls. Kidder's Ferry, rather abund. by the road. May 25-June 10.

317. **S. TERNATUM,** Michx.
 Probably introduced ; local. May 10-June 20.
 Six Mile Cr. "apparently indigenous" in the *Narrows*—"the
middle ravine with *Camptosorus,* detected by J. T. Duncan, May
18, 1870." (*Dr. Jordan.*) Abundant in the ravine of the Ferris
brook where it is probably introduced, as it is along the brook in
the Cemetery.

[**S. telephioides,** Michx.
 Rocky places. "Seneca Lake," *Sartwell in C.:* Prof. *J. Hall
in Flora of N. Y.* I, p. 252, and *S. H. Wright* 22 *N. Y. Rep.* p. 102 ;
the latter writes that his station is just south of Starkey's landing.]

318. **S. TELEPHIUM,** L. LIVE-FOR-EVER.
 Roadsides and fields ; not uncommon. Aug-Sept.

Thoroughly established along the rocky banks of the lake shore, south of L. Ridge and elsewhere ; sometimes becoming a great pest in cultivated fields, east of Ithaca.

DROSERACEÆ.

127. DROSERA, Linn.

319. **D. rotundifolia**, L. ROUND-LEAVED SUNDEW. (**H.** and **C.**)
Sphagnum bogs ; frequent. July 15-30.
Larch Meadow. Freeville sphagnum sw. Malloryville. Round-Marshes and elsewhere.

320. **D. intermedia**, Drev. and Hayne, var. **Americana**, DC. LONG-LEAVED SUNDEW.
"Junius." (*Sartwell in H. and C.*) ; in the Tamarack sw. north of the Pout Pond !

HAMAMELACEÆ.

128. HAMAMELIS, Linn.

321. **H. Virginiana**, L. WITCH HAZEL. (**H.** and **C.**)
Woods and ravines ; common. Oct.-Nov.

HALORAGEÆ.

129. MYRIOPHYLLUM, Vaill.

322. **M. spicatum**, L. COMMON WATER-MILFOIL. (**H.** and **C.**)
Common in the Inlet of Cayuga L., the mouth of Fall Cr., and throughout the lake and outlet in shallow water, to Montezuma. July, Aug.

323. **M. verticillatum**, L. (**C.**)
Scarce ; bayou near mouth of Fall Cr. (*Trelease*, 1878) ! Cayuga Bridge, 1874 ; occasional in the lake-outlet, and in the " Marl-pits." Montezuma Marshes. July-Aug.

324. **M. heterophyllum**, Michx.
Rare ; in the outlet of North Spring at Union Springs. July-Aug.

130. PROSERPINACA, Linn.

325. **P. palustris**, L. MERMAID-WEED. (**C.**)
In water of the larger marshes. Marsh near the Corner-of-the lake (*Mr. Lord.*) ! Cayuga Marshes. Summit Marsh.

131. HIPPURIS, Linn.

326. **H. vulgaris**, L. MARE'S-TAIL.
"Cayuga, *Dr. Jed. Smith*," (*Sartwell in H. and C.*) Not uncommon about the outlet of North Pond at Union Springs 40 or 50 years since. (*Prof'r J. J. Thomas.*) This rare plant—always rare for N. Y.—has not been rediscovered here, although it is possible that it still exists in Cayuga Marshes.

MELASTOMACEÆ.

132. RHEXIA, Linn.

327. * **R. Virginica**, L. DEER-GRASS.
Sandy shores, (not now known in our Flora.) Aug.

In Dr. Thompson's Cat. of Plants near Aurora, (See Regent's Rep. 1841,) as *R. Mariana.* A specimen is in Herb. of Prof'r J. J. Thomas, without locality, but which he thinks was collected near Union Springs. It still exists about Oneida L.

LYTHRACEÆ.

133. LYTHRUM, Linn.

328. L. SALICARIA, L. SPIKED LOOSESTRIFE.
Shores ; rare. July.
Near Inlet south of light-house (*Dr. Jordan*, 1869.) ! Shore of lake west of Inlet, (*F. B. Hine*, 1875.) ! The plants have continued, though not increased.

134. NESEA. Comm. Juss.

329. N. verticillata, H. B. K. SWAMP LOOSESTRIFE. (H. and C.)
Marshes ; not rare. Aug.
Near head of Cayuga L. (*Mr. Lord.*) ! Sheldrake, (*Dr. Jordan.*)! Farley's Point. North of Union Springs. Cayuga Marshes. Junius. Newton's Pond and Pout Pond.

ONAGRACEÆ.

135. EPILOBIUM, Linn.

330. * E. angustifolium, L. WILLOW-HERB. FIRE-WEED. (H. and C.)
Newly-cleared land and lake shore ; common. July-Aug.
Form with pure white flowers at Freeville.

E. HIRSUTUM, L., found in 1874 by Edw. George, near mill west of Case Place.

331. E. palustre, L. var. lineare, Gray. (H. and C.)
Sphagnum and wet marshes ; infrequent. Aug.-Sept.
Larch Meadow. All the sphagnum swamps near Freeville and McLean. Locke Pond. Union Springs and Cayuga Marshes.

332. E. molle, Torr. (C.)
Sphagnum swamps ; infrequent, Aug.
Larch Meadow. Mud Cr., and Malloryville. Round Marshes. Michigan Hollow.

333. E. coloratum, Muhl. (H. and C.)
Wet rocks, ravines and shores ; common. July-Sept.

136. LUDWIGIA, Linn.

334. L. palustris, Ell. FALSE LOOSESTRIFE. (H. and C.)
Low grounds ; common. Aug, Sept.

137. ŒNOTHERA, Linn.

335. * Œ. biennis, L. EVENING-PRIMROSE. (H. and C.)
Low grounds by streams and shores ; common. July-Oct.

336. Œ. biennis, L. var. muricata, Lindl.
Ithaca. Freeville.

337. Œ. biennis, L., var. grandiflora, Lindl. Six Mile Cr. Narrows.
Summit Marsh. Freeville near Fall Cr.

338. **Œ. pumila, L.** DWARF-EVENING-PRIMROSE. **(C.)**
Wet pastures and roadsides ; abundant on the hills. June 20-July.
Not yet found about Ithaca, but frequent near South Hill Marsh :
through Danby, Dryden, Caroline, Enfield, and above Paine's Cr.
Falls.

138. GAURA, Linn.

339. **G. biennis, L.** GAURA. **(H. and C.)**
Sandy shores or dry banks ; occasional. Aug.-Oct.
Near the Armory. Six Mile Cr. Pearson estate. Not rare on
the sandy points all along Cayuga L.

139. CIRCÆA, Tourn.

340. **C. Lutetiana, L.** ENCHANTERS NIGHT-SHADE. **(H. and C.)**
Damp woods ; common. July-Aug.

341. *****C. alpina, L.** **(C.)**
Cool hemlock woods and swamps ; frequent. July-Aug.

CUCURBITACEÆ.

140. SICYOS, Linn.

342. **S. angulatus, L.** STAR-CUCUMBER. **(H. and C.)**
Waste places ; river banks. Aug.-Sept.
Near lake and in town (*Dr. Jordan*)! Negundo Woods. Shurgur's
Glen. Near Ludlowville.

141. ECHINOCYSTIS, Torr. and Gray.

343. **E. lobata, Torr. and Gray.** WILD CUCUMBER. BALSAM APPLE.
Low grounds and river-banks ; not common. Aug.
South of Ithaca, near R. R. Negundo Woods.

FICOIDEÆ.

142. MOLLUGO, Linn.

344. **M. VERTICILLATA, L.** CARPET-WEED.
Roadsides ; scarce. Aug.-Sept.
Near "Nook" green-houses. Near Corner-of-the-lake. On the
Campus, (*Trelease*.)

UMBELLIFERÆ.

143. HYDROCOTYLE, Tourn.

345. **H. Americana, L.** PENNY-WORT. **(H. and C.)**
Springy and moist soil ; common. July-Sept.

144. SANICULA, Tourn.

346. **S. Canadensis, L.** SANICLE. **(C.)**
Ravines and rich woods ; not common. June-July.
Six Mile Cr., Beechwoods flats. Borders of Dryden-Lansing Sw.
Enfield ravine.

347. **S. Marilandica, L.** SANICLE. **(C.)**
Woods and thickets ; frequent. June-July.

145. CONIUM, Linn.

348. **C. MACULATUM, L.** POISON-HEMLOCK. **(H. and C.)**
Waste places ; common ; (poisonous). July-Oct.

[BUPLEURUM ROTUNDIFOLIUM, L. is in the Sartwell Herb. from Gorham, Ontario Co.]

146. CICUTA, Linn.

349. *C. maculata, I.. WATER-HEMLOCK.   (H. and C.)
Marshes and swamps ; frequent. July-Sept.

Too much caution cannot be used in dealing with the Umbellifers. A death occurred in Ithaca a few years since from eating what were reported to be the seeds of this species. They were mistaken for those of some of the harmless members of the Order.

350. C. bulbifera, I..   (H. and C.)
Marshes ; not common. Aug-Sept.

Inlet marshes, and near Renwick. Fall Cr. Beebe Pond. Summit Marsh. Round Marsh. Cayuga Marshes, frequent.

147. CARUM, Linn.

351. C. CARUI, L. CARAWAY.
Roadsides ; not uncommon. July.

Often found in the hill towns, and the pink-flowered form near road from Danby to West Danby.

148. SIUM, Linn.

352. S. cicutæfolium, Gmelin. (*S. lineare*, Michx.)   (H. and C.)
Marshes and swamps ; frequent. Aug.-Sept.

353. S. cicutæfolium, Gm. var. (*S. lineare*, Mx. var. *intermedium*, Torr. and Gray.)
Along the flood-plain of Neguæna Cr., in ditches ; a form with rather broadly ovate leaflets.

149. PIMPINELLA, Linn.

354. P. integerrima, Benth. and Hook. (*Zizia integerrima*, DC. Gray's Man.)   (H. and C.)
Sides of ravines, thickets ; common. May 20-June.

150. CRYPTOTÆNIA, DC.

355. C. Canadensis, DC. HONEWORT.   (H. and C.)
Moist woods ; frequent. June-July.
Casc. Woods and elsewhere.

151. OSMORRHIZA, Raf.

356. *O. longistylis, DC. SMOOTH SWEET-CICELY.   (H. and C.)
Ravines and thickets ; not common. May 20-June.
Negundo Woods. Renwick-Farm slope and elsewhere.

357. O. brevistylis, DC. HAIRY SWEET-CICELY.   (H. and C.)
Rich woods and ravines ; common. May 20-June.

152. CHÆROPHYLLUM, Linn.

358. C. procumbens, Lam. WILD CHEVRIL.
Rich woods ; rare. May.
Negundo Woods, 1874. (*Theo. L. Mead.*)! By Creek near Esty's Tannery. By R. R. near Esty's T.

153. THASPIUM, Nutt.

359. **T. aureum**, Nutt. (H. and C.)
Type form rare. Near cult. field north of Forest-Home. May
20-June.

360. **T. aureum**, Nutt. var. **apterum**, Gray. (H.)
Open meadows and thickets ; frequent.
Near Negundo Woods. Green-Tree Falls. Meadows E. of Etna.
E. of Levanna. Venice.

361. **T. trifoliatum**, Gray, var. **apterum**, Gr. (H. and C.)
Dry woods and open thickets ; common. May 20-June.
The form with winged fruit has not been found.

154. SELINUM, Linn.

362. **S. Canadense**, Michx. (*Coniosclinum canadense*, T. and Gr.)
HEMLOCK-PARSLEY. (C. by Dr. Wright.)
Not rare ; on dripping shaded cliffs in all larger glens. Aug.-
Sept.
Also in cold swamps and marshes, as Fir-tree Swamp, Freeville ;
and along Mud Cr. Beaver Cr. and Round Marsh. Michigan Hol-
low. The swamp habitat is the usual one given in manuals and
catalogues, but Macoun mentions it as on the sea-cliffs of Gaspé.
(*Cat. of Canad. Pl.*)

155. LEVISTICUM, Koch.

363. **L. OFFICINALE**, Koch. LOVAGE.
Roadsides ; not common. Aug.
Near McKinney's. Beyond Geer's Gulf. Enfield. Kings Ferry.
Newfield, toward Saxon Hill.

156. ARCHANGELICA, Hoffm.

364. **A. hirsuta**, Torr. and Gray. (H. and C.)
Dry woods, especially on the hills ; frequent but scattered. Aug.
Case. Woods and near all the ravines. Dart Woods. Turkey Hill.
Pinnacles in Caroline and Danby. Newfield Hills.

365. **A. atropurpurea**,, Hoffm. GREAT ANGELICA. (C.)
Marshes and lake-shore ; frequent. June 15-July.
Indian Spr. Marsh. South of Ithaca. Mud Cr. Beaver Cr.
Cayuga Lake shore, often in sand.

157. PEUCEDANUM, Linn.

366. **P. SATIVUM**, Benth. and Hook. (*Pastinaca sativa L.* WILD
PARSNIP. (C.)
Waste places and along streams ; common. June-Sept.

158. HERACLEUM, Linn.

367. **H. lanatum**, Michx. COW-PARSNIP. (H. and C.)
Alluvial soil ; not common. June, July.
Indian Spr. Marsh. South of Ithaca, not uncommon up the Ne-
guæna valley to Newfield. Occasional along the lake shore.

159. DAUCUS, Tourn.

368. **D. CAROTA**, L. WILD CARROT. (C.)
Dry fields, and creek beds ; becoming common. June-Sept.

Heustis St. Six Mile Creek valley, very common. Turkey Hill, Lansing, and common on sandy shores of the lake.

ARALIACEÆ.

160. ARALIA.

A. SPINOSA, L. HERCULES CLUB. Occasionally spreads from the roots, beyond yards where it is cultivated ; as near Judd's Falls. Aug.

369. * **A. racemosa**, L. SPIKENARD. WHITE-ROOT. (**H.** and **C.**)
In ravines and shaded hollows ; frequent. Aug.

370. **A. hispida**, Michx. BRISTLY SARSAPARILLA.
(**C.** by Dr. Wright.)
Dry banks ; infrequent. June 20–July 20.
Six Mile Cr. south of Green Tree Falls. W. Dryden. Mallory-ville. South of Woodwardia Sw. North of Summit Marsh. Near Montezuma Marshes, (*Dr. S. H. Wright.*)

371. **A. nudicaulis**, L. WILD SARSAPARILLA. (**H.** and **C.**)
Wooded banks ; common. May.
Specimens from South Hill, 1883, (*F. T. Wilson,*) show many ad-ventive, pedicelled flowers at the forks of the corymb, amounting in one case to a full sessile umbel.

372. * **A. quinquefolia**, Gray. GINSENG. (**H.** and **C.**)
Rich woods, scattered ; but a few plants exist in almost every extensive woods, or ravine. June 20–July.
Fall Cr., below the "Primrose Cliffs." Six Mile Cr. Ravines of the lake-shore, and rich woods in Danby, Caroline and Newfield, and about Freeville and McLean.

373. * **A. trifolia**, Gray. DWARF-GINSENG. (**H.** and **C.**)
Rich woods and thickets ; frequent. May.

CORNACEÆ.

161. CORNUS, Tourn.

374. **C. Canadensis**, L. BUNCH-BERRY. (**H.** and **C.**)
Cold woods and borders of sphagnum swamps ; common. June–July.
Fall Cr., north of the Physical Laboratory, a few. West of Varna. Common in Dart Woods, Ellis Hollow, and the Freeville and Danby swamps. A specimen from Freeville, 1880, shows a partially com-pound umbel, groups of flowers being on short peduncles and sur-rounded by partial involucres.

375. * **C. florida**, L. FLOWERING-DOGWOOD. (**H.** and **C.**)
Woods ; frequent. May.

376. **C. circinata**, L'Her. ROUND-LEAVED D. (**H.** and **C.**)
Ravines and high hills ; frequent. June 15–30.

377. **C. sericea**, L. SILKY D. (**H.** and **C.**)
Low grounds ; common. June 12–30.

378. **C. stolonifera**, Michx. RED OSIER. (**H.** and **C.**)
Low grounds ; common. May 25–June 20.

This species and the preceding occasionally bloom a second time, in Aug. or Sept.

379. **C. paniculata**, L'Her.   (H. and **C**.
Hillsides and near marshes ; abundant. June 25-July.

380. **C. alternifolia**, L.   (H. and **C**.)
Ravines and woods ; common. June.

162. NYSSA, L.

381. **N. multiflora**, Wang. PEPPERIDGE.   (H. and **C**.)
Woods and borders of the great marshes ; scarce. June.
Small trees near Fall Cr., north of Beebe Pond. Six Mile Cr., north of Well Falls. South Hill toward Caroline. A few large trees south of Coy Glen, and the Hardenburg Gulf ; also on west shore of Cayuga Lake ; the east shore in Lansing, and on the west side of Cayuga and Montezuma Marshes.

CAPRIFOLIACEÆ.

163. SAMBUCUS, Tourn.

382. **S. racemosa**, L. (*S. pubens*, Mx.)RED-BERRIED ELDER.
                                                      (H. and **C**.)
Ravines and gravelly banks in the hill-towns ; frequent. May.

383. * **S. Canadensis**, L. COMMON ELDER.   (H. and **C**.)
Fields and hedges ; frequent. July 1-20.

164. VIBURNUM, Linn.

384. **V. lantanoides**, Michx. HOBBLE-BUSH.   (C. by Dr. Wright.)
Cold woods and swamps, where it is common. May 10-30.
Fall Creek, west of the Cold-spring, and of the Primrose cliffs Six Mile Creek, and other ravines. Very abundant in the remote swamps.

385. * **V. Opulus**, L. CRANBERRY-TREE.   (H. and **C**.)
Indigenous in deep swamps ; scarce. June.
Indian Spring Marsh. Round Marshes. Fir-tree Swamp at Free-ville. Michigan Hollow. Cayuta Lake Swamp. Probably intro-duced near the Valentine Brook, north of Pleasant Grove Brook, and on Lockwood's Flats.

386. **V. acerifolium**, L., MAPLE-LEAVED ARROW-WOOD. (H. and **C**.
Dry woods, chiefly ; common. June.

387. **V. pubescens**, Pursh. DOWNY ARROW-WOOD.   (C.)
Rocky banks ; frequent. June 5-15.
South of Morrill Hall. Fall Cr., and all ravines. Especially abundant on the lake shore north of King's Ferry.

388. **V. dentatum**, L. ARROW-WOOD.
Low grounds ; common. June.

389. **V. cassinoides**, L. (*V. nudum*, L. *var. cassinoides*. Gray's Man., p. 206.) WITHE-ROD.   (C. by Dr. Wright.)
Sphagnum swamps, where it is frequent. June.
Larch Meadow. South Hill Marsh. Locke Pond, and all the marshes of Dryden and Groton and Danby.

390. **V. Lentago**, L. SWEET VIBURNUM. (H. and C.)
Ravines and low grounds; frequent. May 15-30.

165. TRIOSTEUM, Linn.

391. *T. perfoliatum*, L. FEVER-WORT. (H. and C.)
Rocky open pastures and hedges; frequent. June.
North of Fall Cr. South of Six Mile Creek mill-pond, and else-
where.

166. LINNÆA, Gronov.

392. **L. borealis**, Gronov. TWIN-FLOWER. LINNÆA. (C.)
Banks, and borders of marshes under Pines or Hemlocks. June.
Sparingly in Cascadilla Woods; north of Forest Home; near
Buttermilk Falls; Ludlowville, (*Mr. Lord*); south banks of Taug-
hannock, and Trumansburg ravines; near summit of Saxon Hill;
Fir-tree swamp, S. E. of Danby; island in Summit Marsh, and E.
of Freeville sphagnum bog. Ledyard (*Herb. J. J. Thomas.*)

167. SYMPHORICARPUS, Dill.

393. **S. racemosus**, Michx. SNOW-BERRY. (C.)
Ravines, rocky places and roadsides; infrequent. July.
Six Mile Cr. Enfield ravine. Cayuga Lake shore, etc., possibly
adventive.

394. **S. racemosus**, Mx., var. **pauciflorus**, Robbins.
Rocky banks of large ravines and Cayuga L., where it is abundant. June 20-July 20.
Clearly an indigenous form. Fall Cr., below Ithaca Fall. Taughannock ravine. Cayuga L., rather common, especially on the
east-shore cliffs from Mc'Kinney's to Levanna.

[S. VULGARIS, Michx. INDIAN CURRANT. Borders of Seneca Lake,
(*Sartwell in Herb.*), " probably introduced,"—(*S. H. Wright.*)

168. LONICERA, Linn.

395. L. XYLOSTEUM, L. FLY-HONEYSUCKLE (OF EUROPE.)
South Hill by the R. R., near North Quarry, 1880, (*F. C. Curtice*)!
Also south of the Quarry, a quarter of a mile from the first station.
1884.

396. **L. oblongifolia**, Muhl. SWAMP FLY-HONEYSUCKLE. (H. and C.)
Deep swamps; rare. June 1-20.
Sparingly in Michigan Hollow swamp, near the *Rhododendron;*
Sw. at the head of Locke Pond. (*Dr. Ch. Atwood.*) (Otter Cr.,
Cortland.) Lowery's pond, Junius. (Tamarack Sw., west of Sa-
vannah.)

397. *L. ciliata*, Muhl. FLY-HONEYSUCKLE. (H. and C.)
Ravines and damp woods; common. May.

398. L. TARTARICA, L. TARTARIAN HONEYSUCKLE.
Pastures and banks; becoming frequent. May 25 June 15.
South Hill, near the "Incline" and R. R.; very abund. Also in
Casc. Cr., Fall Cr., and frequent on shore of Cayuga L. On Fron-
tenac Island are specimens with white flowers and salmon-colored
berries.

399. L. SEMPERVIRENS, L.

South Hill, beyond the "Incline," (1883-85), borders of a thicket in a rocky pasture ; probably introduced. July, Aug.

400. L. hirsuta, Eaton. HAIRY HONEYSUCKLE. (H. and C.)
Moist copses ; rare. June.

Freeville, west of the Southern Cent. R. R. North of Summit M. (Woods south of Spencer Sta., Tioga Co.) Our forms are all less hairy than the more northern ones.

401. L. glauca, Hill. (L. parviflora, Lam., Man., p. 204.) SMALL HONEYSUCKLE. (H. and C.)

Rocky banks and slops ; frequent. May 10-June 10.

A form occurs sparingly on the High Cliffs north of King's Ferry bearing pure yellow flowers, but otherwise like the type.

402. A form of (401)—possibly a *variety*—occurs in the old woods on our higher and more remote hills, having leaves mostly distinct excepting the upper pair, green and smooth above, underneath less glaucous than in the type, and more or less hairy. The ovaries and peduncles are somewhat glandular when young, when mature only slightly hairy, and in color orange-red. Flowers not yet seen. Possibly this formed a part of the old var. *Douglassii* of *L. parviflora.*

169. DIERVILLA, Tourn.

403. D. trifida, Moench. BUSH HONEYSUCKLE.
Dry woods, and rocky banks of ravines ; frequent. June.

RUBIACEÆ.

170. HOUSTONIA, Linn.

404. H. cærulea, L. BLUETS. INNOCENCE. WILD-FORGET-ME-NOT.
(H. and C.)
Damp meadows and pastures ; local. May-Oct.

South Hill, extending ½ to 1 mile in every direction from South Hill Marsh.

H. PURPUREA, L., var. LONGIFOLIA, Gray, appeared on the Fiske-McGraw grounds, 1884. (C. by Dr. Wright.)

171. CEPHALANTHUS, Linn.

405. *C. occidentalis, L. (H. and C.)
Swamps ; frequent. Indian Spring Marsh, etc. Aug.

172. MITCHELLA, Linn.

406. *M. repens, L. PARTRIDGE-BERRY. SQUAW-PLUM. ONE-BERRY. (H. and C.)
Woods ; common. June 25-July 25.

Form with leafy berries, Case. Cr. (*Mr. Lord,* 1880.) 1881! near Brookton, 1881. (for figures of these specimens, see *Bulletin Torr. Botan. Club,* Vol. X, pl. XXVI. Jan. 1883.) Form with white berries, Round Marsh woods, 1883. (O. E. Pearce.)

173. GALIUM, Linn.

407. G. Aparine, L. CLEAVERS. (C.)
Ravines and damp thickets ; abundant. May.

408. **G. pilosum**, Ait.

    Dry banks; rare. July.

    Wild bank in the rear of the Valley Cemetery. Junius, near ponds, and S. W.; also in *Sartwell Herb.* from "Junius."

409. **G. circæzans**, Michx.   (H. and **C.**)

    Woods; frequent. July.

410. **G. lanceolatum**, Torr.   (H. and **C.**)

    Woods and ravines; frequent. June–July.

411. **G. boreale**, L. NORTHERN BEDSTRAW.   (H. and **C.**)

    Rocky banks and slopes; common. June–July.

    Particularly abundant, South Hill and the lake shore.

412. **G. trifidum**, L. SMALL B.   (H. and **C.**)

    Wet places and marshes. Near Cayuga L., etc.; common. June–July.

413. **G. trifidum**, L. var. **pusillum**, Gray.   (H.)

    Sphagnum marshes and on Ithaca marsh; frequent.

414. **G. trifidum**, L. var. **latifolium**, Torr.   (H.)

    Marsh near the lake; near Kings Ferry; scarce.

415. **G. asprellum**, Michx. ROUGHER B.   (H. and **C.**)

    Swamps; common. July–Oct.

416. **G. triflorum**, Michx. SWEET B.   (H. and **C.**)

    Woods; frequent. July.

VALERIANACEÆ.

174. VALERIANA, Tourn.

417. **V. OFFICINALIS**, L. GARDEN VALERIAN.

    Escaped, near Locke Pond; and north of Taughannock ravine in woods where it has bloomed many years. June.

418. **V. sylvatica**, Banks. Sphagnum sw., June; in *Herb. of J. J. Thomas*, from Junius; probably 1827. The usual statement is that this rare plant was first detected in New York, in Savannah, Wayne Co., by Dr. Sartwell, 1833. (See *Gray in New and Rare Pl. of N. Y.*, p. 226.) The only other N. Y. stations as yet reported are: north of Newark, Wayne Co., (*Mr. Hankenson*); W. Bergen Sw.; Warren, Herkimer Co.; and Pine Plains, Dutchess Co.

175. VALERIANELLA, Tourn.

419. **V. OLITORIA**, Poll. (*Fedia olitoria*, Vahl.) CORN SALAD.

    On Frontenac Id. Cayuga L. (*Mrs. Prof. S. J. Brun, in* 1882)! Rare. May.

[**V. chenopodifolia**, DC. (*Fedia Fagopyrum*, T. and Gr.) is in *Herb. Sartwell*, from Penn Yan.]

DIPSACEÆ.

176. DIPSACUS, Tourn.

420. *** D. SYLVESTRIS**, Mill. TEASEL.

    Roadsides and fields; common. July–Oct.

177. SCABIOSA, Linn.

421. **S. AUSTRALIS**, Wulf. SCABIOUS.

    Wet, often marshy places; introduced and local. Aug–Oct.

Near Cayuga Lake from Union Springs south to Farley's Point and Lockwood's Flats. Marsh north of village; and on Frontenac Id.; abundant within the limits. *Scabiosa* differs from *Dipsacus*, chiefly in the scales of the receptacle, which are either not prickly or reduced to hairs.

S. AUSTRALIS, Wulf., is a perennial herb, ½–¾ meter in height, with few opposite branches forming a corymb. Radical leaves spatulate; cauline opposite, lower oblanceolate and tapering into a petiole, upper linear-lanceolate and sessile, entire, obtuse or acute, glabrous or slightly hairy on the veins; light green. Heads on long peduncles globular or oblong (1½ cm.), flowers small, cerulean blue. Involucel, glabrous, 4-lobed; *lobes short, obtuse; calyx-lobes not extending into bristles.* Detected Aug. 1881. The history of its introduction (in this the only station in America so far as I know), is clearer than in most such cases; for in the Herb. of Prof'r J. J. Thomas are specimens taken from his father's garden, where it was cultivated fifty years ago. The garden was back from the lake only a mile or two.

COMPOSITÆ.

178. MIKANIA, Willd.

422. **M. scandens**, Willd.

In thickets, and on wild grasses; only along the Cayuga Marshes, where it is frequent. Aug.

179? EUPATORIUM, Tourn.

423. *** E. purpureum**, L. TRUMPET-WEED. PURPLE THOROUGH-
WORT. (H. and C.)

Low grounds; one of the commonest weeds. Aug.–Sept.

424. **E. sessilifolium**, L. (C.)

Declivities of ravines and lake shores; scarce. Aug.–Sept.

Fall Cr. below Ithaca Fall. Near the "Nook." Cayuga L. South of Shurgur's Glen, and north of Myers Pt. Enfield ravine.

425. *** E. perfoliatum**, L. BONESET. THOROUGHWORT. (H. and C.)

Low grounds; common. Aug.–Sept.

426. *** E. ageratoides**, L. f. WHITE SNAKE-ROOT. (H. and C.)

Ravines and woods in shade; abundant. Aug.–Sept.

[**Liatris scariosa** is in *Sartwell's Cat.*, 1844.]

180. SOLIDAGO, Linn. GOLDEN ROD.

427. **S. squarrosa**, Muhl. (H. and C.)

Dry woods and banks; frequent. Sept.

All our ravines, and especially abundant on banks of Cayuga L.

428. **S. cæsia**, L. (H. and C.)

Woods; common, (the *var. axillaris*, Gray, frequent.) Sept.

429. **S. latifolia**, L. (H. and C.)

Sides of ravines, shaded wet places; not uncommon. Sept.

430. **S. bicolor**, L. (H. and C.)

Fields and woods; common. Sept.

[? **S. odora**, Ait.; "rocky banks of Seneca L. *Vasey.*" See Paine's C. p. 94.]

131. **S. uliginosa**, Nutt. (*S. stricta* of Gray's Man., p. 240.)

Sphagnum swamps; where it is frequent. Sept.

The strict form occurring at Larch Meadow and the Round Marshes. A more thyrsoid form in Round Marshes and along Mud Cr. Some of the latter can scarcely be distinguished from *S. neglecta* except the racemes are not secund.

132. **S. patula**, Muhl. (**C.**)

Wet places and marshes; common. Sept.

133. **S. rugosa**, Mill. (*S. altissima*, L. of Gray's Man., p. 243.)

(**H. and C.**)

Borders of woods and thickets; frequent. Aug.-Sept.

Cascadilla Woods near the Armory. South Hill. Freeville and elsewhere.

134. **S. ulmifolia**, Muhl. (**H.**)

Ravines and declivities; scarce. Sept.

Fall Cr. Near sand-bank north of Fall Cr. Mills. Somewhat abund. north of Myers Point. South of Willet's Sta. Big Gully.

135. **S. neglecta**, Torr. and Gray. H. and **C.**)

Sphagnum swamps; scarce. Sept.

Larch Meadow. Round Marshes.

136. **S. neglecta**, T. and Gr. var. linoides, Gray. (*S. linoides*, Sol. Man., p. 243.)

Newton's Ponds. Lowery's Ponds, Junius. (Same station by *Sartwell, H. and C.*)

137. **S. arguta**, Ait. (The *Solidago Muhlenbergii*, T. and Gr. of Man., p. 243.) (**H. and C.**)

Sides of ravines and along the lake; frequent. Aug.-Sept.

In sphagnum swamps it becomes variable, its leaves often roughish above.

138. **S. juncea**, Ait. (The *S. arguta* of Gray's Man., p. 243.)

(**H. and C.**)

Banks, borders of woods; common. July 20-Aug.

The handsomest and one of the earliest Golden-rods. The narrow leaved form along the lake blooms earliest. Occasionally it grows and blooms in the crevices of the Tully limestone, at the height of 8-12 cm.

139. **S. serotina**, Ait. (*Solidago gigantea*, Ait, of Man., p. 245.)

(**H. and C.**)

Borders of woods; common. An early bloomer. July 15-Aug.

140. **S. serotina**, Ait., var. gigantea, Gray, (the *S. serotina* of Man., p. 245.) (**H. and C.**)

Often on marshes, also damp fields; frequent. Aug.

South Hill. Inlet Marsh. Summit Marsh. Near Black Lake. Dryden L. valley and elsewhere.

141. **S. Canadensis**, L. (**H. and C.**).

Fields and thickets, etc.; very common and late blooming. Sept.-Oct.

Below Sheldrake are specimens with smooth leaves.

442. **S. nemoralis**, Ait.   (**H.** and **C.**)
    Fields; common. Sept.

443. **S. Ohioensis**, Riddell.
    Sphagnum marsh, about Newton's and Lowery's Ponds, Junius.
    Aug. 20–Sept. 20.
    "Junius," (*Sartwell, H. and C.*) Somewhat abund. at this sta-
    tion. Probably West Bergen Sw. is the only other station in the
    State.

444. **S. lanceolata**, L.   (**C.** by Dr. Wright.)
    Low places in fields; scarce. Aug.–Sept.
    South Hill south of S. S. 420. White Church. Near Summit
    Marsh. Dryden Lake. Mud Cr. Locke Pond. Groton. (*F. C. C.
    and F. L. K.*) Not yet found on the western hills of our Flora.

BELLIS, Tourn.

B. PERENNIS, L. ENGLISH DAISY. Spontaneous on Mr. Lord's lawn
    several years.

181. SERICOCARPUS, Nees.

445. **S. conyzoides**, Nees. WHITE-TOPPED-ASTER.   (**H.** and **C.**)
    Dry woods near ravines, etc.; not uncommon. Aug.

182. ASTER, Linn. ASTER.

446. **A. corymbosus**, Ait.   (**H.** and **C.**)
    Hedges and thickets; frequent. July–Aug.

447. **A. macrophyllus**,   (**H.** and **C.**)
    Along brooks and in rich woods; abundant. Aug.–Sept.

448. **A. Novæ Angliæ**, L. NEW ENGLAND ASTER.   (**H** and **C.**)
    Hillsides and along streams; a very abundant, late bloomer.
    Sept. 12–Oct.

449. **A. Novæ Angliæ**, L. var. **roseus**, Gr.
    Abund. on South Hill. Near Case. Cr. and Union Springs; a form
    with light pink flowers at Union Springs.

450. *var.* with light blue flowers at Union Springs. South of Levan-
    na. Near Osmun's Sta.

451. **A. undulatus**, L.   (**H.** and **C.**)
    Woods; common. Sept.–Oct.

452. **A. cordifolius**, L.   (**H.** and **C.**)
    Woods; common. Sept.–Oct.

453. **A. sagittifolius**, Willd.   (**H.** and **C.**)
    Banks of Cayuga L. Sept.–Nov.
    This species is abundant from Esty Glen to Union Springs, entire-
    ly replacing in some sections, *A. cordifolius*. Very rarely found
    away from the shore.

454. **A. lævis**, L.   (**C.**)
    Woods and especially abund. on lake shore; common. Sept.–
    Nov.
    Sometimes with very glaucous leaves, (the typical form); some-
    times leaves bright green.

455. A. ERICOIDES. L.
    Dry soil; local, probably introduced here. Sept.–Oct.

By R. R. east of Freeville. By R. R. south of W. Danby.

456. **A. vimineus**, Lam. (*A. Tradescanti* of Man., p. 232, for most part.)

Low grounds and woods as Casc. Cr. Fall Cr. and Freeville; frequent. Aug.–Sept.

457. **A. diffusus**, Ait. (*A. miser*, L. of Man., for most part.)
(**H. and C.**)

Woods. South Hill. Freeville; type not very common. Aug.–Sept.

458. **A. diffusus**, Ait. var. **thyrsoideus**, Gray; occurs near shore of Cayuga L. South of Lake Ridge and in woods on Utts Point.

459. **A. diffusus**, Ait. var. **hirsuticaulis**, Gray; (believed by Dr. Gray to be nearest this,) occurs in Fall Creek and Freeville, while a more pronounced form grows at Mud Cr.

460. **A. Tradescanti**, L. (*A tenuifolius*, of Man., p. 233 in part.)
Cayuga L. shore; not common. Sept.–Oct.
Farley's Point. Near U. Springs depot. Lake Ridge Pt.

461. **A. paniculatus**, Lam. (*A. carneus*, Nees and *A. simplex*, Willd, of Man., p. 233, mainly.) (**H. and C.**)
Low grounds; very common. Sept.–Oct.

The forms with large whitish or bluish-white heads, dark-green thin, serrate leaves and widely branching stems seem to be the typical ones. It is excessively variable but one of the commonest species in this region.

462. **A. junceus**, Ait. (*A. æstivus* of Man., p. 233 mainly.)
Sphagnum swamps; infrequent. July 20–Sept. 15.

A handsome *Aster*, in the Round Marshes and open bogs in Gracie's Sw. Near "Marl-Ponds." Fleming Meadow, and Cayuta Lake.

463. **A. Novi-Belgii**, L. ?
Fleming Meadow.

464. **A. prenanthoides**, Muhl. (**H. and C.**)
Wet places and swamps; common. Sept.–Oct.

465. **A. puniceus**, L. (**H. and C.**)
Marshes and along streams; common. Sept.–Oct.
Form with white flowers, near Forest-Home, (*Dr. Jordan.*)

466. **A. puniceus**, L. var. **lævicaulis**, Gray.
A well-marked form near R. R. north of Ludlowville, and near canal E. of Montezuma.

467. **A. puniceus**, L. *var.* with purple stems, nearly smooth, and flowers pinkish.
Near Union Springs, Round Marsh, Freeville and Mud Creek.

468. **A. umbellatus**, Mill. (*Diplopappus umbellatus*, Torr. and Gr.)
(**H. and C.**)
Ravines and near sphagnum swamps; abund. Aug.–Sept.

469. **A. acuminatus,** Michx. (H. and C.)
Cold elevated woods and ravines usually. Aug.–Sept.
Six Mile Cr., and woods and ravines in Newfield, Danby, Caroline
and Dryden ; only common in tracts recently cleared.

182². ERIGERON, Linn.

470. **E. bellidifolius,** Muhl. ROBINS-PLANTAIN. (H. and C.)
Grassy banks ; frequent. May 15–June 2.

471. **E. Philadelphicus,** L. PINK FLEABANE. (H. and C.)
Grassy banks ; often on wet rocks of ravines : frequent. June.

472. **E. annuus,** Pers. TALL-DAISY. (H. and C.)
Fields ; common. July–Nov.

473. **E. strigosus,** Muhl. DAISY FLEABANE. (H. and C).
Field ; common. July–Nov.
Near Union Springs, Round-Marsh, Freeville and Mud Creek.

174. **E. Canadensis,** L. HORSEWEED. (H. and C.)
Fields and waste places ; common. Aug.–Nov.

183. ANTENNARIA, Gærtn.

475. **A. plantaginifolia,** Hook. PLANTAIN-LEAVED EVERLASTING.
(H. and C.)
Dry banks, and wet sandy fields ; common. April–May.

184. ANAPHALIS, DC.

476. **A. margaritacea,** Benth. and Hook. (*Antennaria margaritacea,*
R. Br.) PEARLY EVERLASTING. (H and C.)
Near ravines, and fields on high hills where it is common. Aug.–
Sept.

185. GNAPHALIUM, Linn.

477. **G. polycephalum,** Michx. (H and C.)
Dry woods ; not common. Sept.
Case. Woods. Fall Cr. Near Ludlowville. W. Danby, on the
pinnacles. Enfield ravine.

478. **G. decurrens,** Ives. COMMON EVERLASTING. (H. and C.)
Woods and old fields ; frequent. Aug.–Sept.

479. ***G. uliginosum,** L. CUDWEED. (H. and C.)
Wet places and shores ; common. July–Oct.
Specimens from Cortland Marl-Ponds, less than 1 cm. tall, but
fruiting.

[Gnaphalium purpurem, L., in Sartwell's Herb. and Cat., from Penn
Yan.]

186. INULA, Linn.

480. *I. HELENIUM, L. ELECAMPANE. (H. and C.)
Pastures and along streams; frequent. July–Aug.

187. POLYMNIA, Linn.

481. **P. Canadensis,** L. LEAF-CUP. (H. and C.)
Rich or rocky soil ; infrequent. Aug.
Negundo Woods. Ravine of Ferris Brook. North Pinnacle in
Caroline. Cedar swamp in Tyre.

452. P. Uvedalia, L..   (H. and C.)
    Ravines ; rare. Aug.
    Salmon Cr. ravine, below the spring.   Big Gully, above the Falls.
    (E. of Bellona, *S. H. Wright.*)

188. AMBROSIA, Tourn.

483. A. trifida, L.   GREAT RAG-WEED.   (H. and C.)
    Near streams, in rich soil ; common.   July, Aug.

484. A. artemisæfolia, L.   RAGWEED.   (H. and C.)
    Fields ; very common.   July–Oct.

189. XANTHIUM, Tourn.

485. X. Canadense, Mill.   COCKLE-BUR.   (H. and C.)
    Streams and shores ; common.   Aug.–Oct.

486. X. Canadense, Mill, var. echinatum, Gray.
    Same range ; common.

ZINNIA ELEGANS, L., occasionally escapes ; as on banks of Fall Cr.,
1882 ; and roadside on the marsh.

190. HELIOPSIS, Pers.

487. H. lævis, Pers.   OX-EYE.   (H. and C.)
    Alluvial soil, where it is very abundant.   July 1–Oct.
    Very abundant along the lake, its marshes, and the Negunena val-
    ley ; also Fall Cr. at Freeville.   Six Mile Cr. and elsewhere.

191. RUDBECKIA, Linn.

488. R. hirta, L.   YELLOW DAISY.   OX-EYE DAISY.   CONE-FLOWER.
                                           (C. by Dr. Wright.)
    Fields ; frequent and becoming a weed.   July-Aug.

489. R. laciniata, L.   CONE-FLOWER.   (H. and C.)
    Creeks, marshes and swamps ; frequent.   Aug. 1-Sept. 30.

192. HELIANTHUS, Linn.

H. ANNUUS, L., the great "SUNFLOWER" occasionally appears in
    waste places, but is not permanent.

490. H. divaricatus, L.   (H. and C.)
    Dry banks and woods ; common; rarely with three leaves in a
    whorl. Aug.-Sept.

[H. divaricatus, "var. foliis-ternatis, Torr. and Gray," at Penn Yan,
    (*in Herb. Sartwell.*)]

491. H. strumosus, L.   (H. and C.)
    Woods, or shaded rich soil ; abundant.   Aug.–Sept. 20.
    A stout, broad and hairy leaved form, south of Ithaca and near
    Cayuga Bridge.   A form with narrowly lanceolate leaves, near
    King's Ferry and Lockwood's Flats ; the same in Herb. of J. J.
    Thomas, as *H. tracheliifolius.*

492. H. decapetalus, L.   (H. and C.)
    Dry woods or ravines ; frequent.   Aug. 15–Sept.

493. H. tuberosus, L.   JERUSALEM-ARTICHOKE.   (H. and C.)
    Alluvial ground or shores ; not uncommon.   Sept.

Fall Creek above. Beebe Pond, (apparently indigenous), and Six Mile Cr. The escaped hairy form, near Ithaca and Cayuga L. shore, on Lockwood's Flats and elsewhere.

[H. giganteus, L. at " P. Yan," *Sartwell in H.*]

[Actinomeris squarrosa, Nutt. Borders of Crooked Lake, Yates Co. *Sartwell in Herb.*, also in *Torrey's Flora of N. Y.*, I, p. 384.]

193. COREOPSIS, Linn.

494. C. trichosperma, Michx.
Rare. Specimen in herb., from head of Cayuga Lake, Oct., 1879. (*F. C. Curtice.*)

195. C. discoidea, Torr. and Gray.
Exsiccated places ; scarce. Sept.–Oct.
On old logs, shores of Dryden L. Near Cayuga L., on the Marsh. (Hybrids between this and *Bidens frondosa*, L., have been collected at Cascade, Owasco L.)

194. BIDENS, Tourn.

496. B. frondosa, L. COMMON BEGGAR-TICKS. (H. and C.)
Low grounds and waste places ; common. Aug.-Oct.

497. B. connata, Muhl. (H. and C.)
Along creeks, rich soil ; abundant. Aug.-Sept.

498. B. connata, Muhl., var. comosa, Gray.
Common along Salmon Cr. and Myers P't. ; also, Sheldrake P't.

499. B. cernua, L.
Wet places, shores and bottom of ravines ; frequent. Sept., Oct. At Marl Ponds, Cortland, it fruits at heighth of 7-12 cm. Junius, (*Sartwell in H. and C.*)

500. B. chrysanthemoides, Michx. (H. and C.)
Marshes and ditches ; common. Aug.-Oct.

501. B. bipinnata, L. SPANISH NEEDLES.
Rare as yet. Lake Ridge P't., 1880. " Ovid," (*S. H. W.*) Aug.-Oct.

502. B. Beckii, Torr. WATER MARIGOLD.
Still, shallow ponds; rare. Aug.
Locke Pond, 1881. (Pond at Homer, *Mr. Lord.* Mill-pond at Cortland. *Dr. Ch. Atwood.*) Cayuta L. in N. W. part.

195. HELENIUM, Linn.

503. H. autumnale, L. SNEEZE-WEED. (H. and C.)
Alluvial soil ; frequent. Aug. 20-Oct.
Fall Cr., and elsewhere ; particularly abundant on Cayuga Marshes.

196. ANTHEMIS, Linn.

504. A. COTULA, L. (*Maruta Cotula*, L.) MAY-WEED. (H. and C.)
A common weed. Aug.-November.

505. A. ARVENSIS, L. CORN-CHAMOMILE. (H. and C.)
Fields ; becoming abund. near Ithaca and the lake. Beginning to flower in June.

197. ACHILLÆA, Vaill.

506. *A. millefolium, L. YARROW. (H. and C.)
Fields; common, the pink flowered form at Ball Hill. July-Nov.

198. CHRYSANTHEMUM, Tourn. *Leucanthemum,*Tourn.

507. C. LEUCANTHEMUM, L. DAISY. (H. and C.)
Fields; common. June-Nov.

508. C. PARTHENIUM, Pers. FEVERFEW.
Thoroughly established throughout Stevens Woods on west shore of Cayuga. Occasional near Inlet and in Ithaca.

199. TANACETUM, Tourn.

509. T. VULGARE, L. TANSY. (H. and C.)
Roadsides, (the form "*var. crispum*" is also frequent.) July-Sept.

200. ARTEMISIA, Tourn.

510. * A. ABSINTHIUM, L. COMMON WORMWOOD. (H. and C.)
By roadsides north of Willets Sta. formerly, (1879.) Aug.

511. *A. ABROTANUM, L. SOUTHERN-WOOD.
In W. Junius, by road S. W. of the Pout Pond.

201. TUSSILAGO, Tourn.

512. T. FARFARA, L. COLTS-FOOT. (H. and C.)
Damp, clay soil, and banks of streams; common. April.

202. PETASITES, Tourn.

513. P. palmata, Gray. (*Nardosmia palmata,* Hook.) SWEET COLTS-FOOT.
Near Buttermilk Cr., in a marsh south of "Pulpit Rock", (*G. W. Wood,*)! 1873. It continued till 1876, when it was destroyed by felling the trees which shaded it. Found in 1875 in a swamp near Smith's Corners, north of Cayuta L. This station was also destroyed a few years ago in the same way. ("Dundee, Yates Co." *S. H. Wright, in Herb. Sartwell.* Station is 2 m. E. of Dundee.)

203. SENECIO.

514. * S. aureus, L. GOLDEN RAG-WORT. (H. and C.)
Wet swamps; frequent. May 15-June.

515. S. VULGARIS, L. GROUNDSEL. (H. and C.)
Scarce. "Below Casc. Place, by *Farquahr.,* 1869." (*Dr. Jordan.*) On the Campus, 1881, (*F.L.K.*) It occurs near Edgewood, near the Armory, and on the Fiske-McGraw grounds. Aurora, by the R R.

204. ERECHTHITES, Raf.

516. E. hieracifolia, Raf. FIRE-WEED. (H. and C.)
Woods, fields and recently cleared land; common. Aug.-Sept.

205. ARCTIUM, Linn.

517. * A. LAPPA L. (*Lappa officinalis,* All., Man., p. 275.) BURDOCK. (H. and C.)
A weed; common. Aug.-Nov.

518. A. LAPPA, L. var. MINUS, Gr., is abundant on the lake shore.

206. CNICUS, Tourn. (*Cirsium, DC. of Man.*)

519. C. ARVENSIS, Hoffm. CANADA THISTLE.   (H. and C.)
Fields ; excessively common. July-Oct.

520. C. ARVENSIS, Hoff. *var.* ALBIFLORUS, appears more and more
abundantly each year.

521. *C. LANCEOLATUS, Hoffm. COMMON THISTLE.   (H. and C.)
Fields ; common. Aug.-Oct.

522. C. pumilus, Torr. PASTURE THISTLE.   (H. and C.)
Pastures ; scarce. Aug.-Sept.
North of Fall Cr. South Hill. E. of the South Pinnacle in
Caroline.

523. C. altissimus, Willd., var. discolor, Gray. (*Cirsium discolor,*
Spring.)   (H. and C.)
Univ. Grove, formerly. Near Indian Spring Marsh. Negundo
Woods. North of Levanna and Lockwood's Flats. On banks ;
scarce.

524. C. muticus, Pursh. SWAMP THISTLE.   (H. and C.)
Swamps ; frequent. Aug.-Oct.

[ONOPORDON ACANTHIUM, L. Near Penn Yan, by S. H. Wright.]

[SILYBUM MARIANUM, Gaertn. Also near Penn Yan, by Wright.]

207. ECHINOPS, Linn.

525. E. RITRO, L. GLOBE THISTLE.
Caroline Center, escaped from cult. and spreading rather freely,
along a brook. (*F. T. Wilson*, 1885.)

208. CENTAUREA, Linn.

526. C. CYANUS, L. CORN-FLOWER,   (C.)
Campus. E. State street, Ithaca, and elsewhere, appearing almost
every year. In 1885, in a wheat field east of Case Woods, many
specimens. (*F. T. Wilson*.)!

C. NIGRA, L. On the Campus near the McGraw building, 2875.
(C. by Dr. Wright.)

209. CICHORIUM, Tourn.

527. C. INTYBUS, L. CHICCORY.   (H. and C.)
Fields ; not common. Aug.-Sept.
Has appeared on the Campus at intervals, but is established only
in fields north of Dryden Lake ; near Romulus Sta. ; by road west
of Enfield Falls, and on Crobar Point.

210. TRAGOPOGON, Linn.

528. T. PORRIFOLIUS, L. SALSIFY.
Fields ; becoming frequent. July-Oct.
Campus, south of Sage Coll. Hazen St. Quarry St. South
Hill. Enfield, and elsewhere.

529. T. PRATENSIS, L. YELLOW GOAT'S-BEARD.
South Hill. Near Levanna (1879.) Venice and Genoa, occasional. A weed in farms west of Locke Pond., (*Dr. C. Atwood.*)!

211. HIERACIUM, Tourn.

530. H. AURANTIACUM, L.
Recently introduced. In 1885, a few specimens were found south of Willow Pond race, (June 15); near Sw. on Owasco Hill in Venice; near Caroline Center, (*F. T. Wilson.*)

531. H. Canadense, Michx.   CANADA HAWKWEED.   (H. and C.)
In dry woods; not rare.   Aug.-Sept.
Casc. Woods. Fall Cr., near Cayuga L., and elsewhere.

532. H. paniculatum, L.   (H. and C.)
In woods; frequent.   Aug.-Sept.

533. H. venosum, L.   RATTLESNAKE H.   (H. and C.)
In dry woods; common.   June-July.
Certain forms with leaves cauline and scarcely purple-veined, approach *H. Marianum*, Willd., but I do not think that species is here.

534. H. scabrum, Michx.   ROUGH H.   (H. and C.)
Dry woods, frequent.   Aug.-Sept.

535. *H. Gronovii, L., is in Dr. Thompson's Cat., and is reported from near Auburn by I. H. Hall.   (See *Paine's Cat.*, p. 99.)   It is likely to occur in the sandy woods of Junius.

212. PRENANTHES, Vaill.

536. P. alba, L.   (*Nabalus albus*, Hook).   WHITE-LETTUCE.
(H. and C.)
Woods; abundant.   Aug.-Sept.
A form in Cascadilla Woods has stem leaves deeply 3-5 parted the lobes pinnatifid.

537. P. Serpentaria, Pursh.   (*N. Fraseri*, DC.)   LION'S-FOOT.
Dry woods; not uncommon.   Aug.-Sept.
It occurs in the dry woods along all our ravines, and the pinnacles of Danby and Caroline.

538. P. altissima, L.   (*N. altissimus*, Hook.)   (H. and C.)
Moist shaded soil; frequent.   Aug.

213. TARAXACUM, Haller.

539. *T. officinale, Web.   (*T. Dens-leonis*, Desf.)   DANDELION.
(H. and C.)
Fields; very common.   Blooming from May till Dec.

214. LACTUCA, Tourn.

540. L. SCARIOLA, L.   PRICKLY-LETTUCE.   (H. and C.)
Gravelly field on Trumansburg Point. Cayuga L. A few rods south of the landing, where it is abundant.   Aug.

541. L. Canadensis, L.   WILD LETTUCE.   (H. and C.)
Rich soil and Cayuga L. shore; abundant.   July-Sept.

542. L. integrifolia, Bigelow.   (*L. Canadensis*, L. var. *integrifolia*, T. and Gr.)   (C. by Dr. Wright.)
Drier soil; not common.   Aug.
Brookton. Cayuga L. at Norton's. Union Springs and Cayuga Bridge.

543. **L. leucophæa,** Gray. (*Mulgedium leucophæum,* DC. Man., p. 282.) BLUE-LETTUCE.   (**H.** and **C.**)
Low grounds ; common. Aug.-Sept.

L. SATIVA, L. (the GARDEN LETTUCE) often appears for a season along roadsides near Ithaca.

215. SONCHUS, Tourn.

544. * S. OLERACEUS, L. SOW-THISTLE.   (**H.** and **C.**)
Waste places and roadsides ; abund. Aug.-Sept.

545. S. ASPER, Vill. SPRING SOW-THISTLE.   (**H.** and **C.**)
Roadsides and especially frequent on Cayuga L. Aug.-Sept.

546. S. ARVENSIS, L. FIELD SOW-THISTLE.
Meadows and shores ; frequent. July.-Sept.
The Mack Farm, south of Ithaca. Eddy St. Case. St., Ithaca. Myers Pt. Sheldrake. Union Spr. and Cayuga. First reported from N. Y. by *Hon. H. B. Lord,* "shores *of Cayuga Lake,*" in 4th Ed. Gray's Man., (1863). Also from the same station, with "*Rochester, G. W. C.,*" and "*Staten Id. T. F. A.,*" added, in 18th N. Y. Rep. p. 205, (1865.) Auburn, *I.H.Hall,* Paine's Cat. (1864). Though the flowers are handsome and fragrant, the strong perennial rootstock is a pest in cultivated soils.

LOBELIACEÆ.

216. LOBELIA, Linn.

547. * **L. cardinalis,** L. CARDINAL-FLOWER.   (**H.** and **C.**)
Low meadows and borders of ponds ; frequent. July 20.-Sept.
Lake marshes near Inlet and Fall Cr. Case. Cr. upper valley. White Church valley and elsewhere. Especially abundant about some of the Cortland Marl Ponds. Summit Marsh, and Cayuga Marshes.

548. * **L. syphilitica,** L. GREAT BLUE-LOBELIA.   (**H.** and **C.**)
Wet places ; abundant, Aug.-Sept.
A form with white flowers near Mud Cr., Freeville, (and near Penn Yan, *Sartwell Herb.*)

549. * **L. Kalmii,** L.
Wet rocks and cliffs, in ravines ; where it is frequent. July.-Sept. Fall Cr. and Case. Cr. (*Dr. Jordan*)! Six Mile Cr. Buttermilk Cr. Enfield ravine. Taughannock ravine, Salmon Cr. ravine. A stout strict form, often with bractlets near the base of short pedicels, occurs on open marshes and limestone shores, as Larch Meadow, and Farley's Point. (see Paine's Cat. p. 100, for apparently same form.) "Junius" (*Sartwell H. and C.*)!

550. * **L. inflata,** L. INDIAN TOBACCO.   (**H.** and **C.**)
Fields ; common. July Sept.

CAMPANULACEÆ.

217. SPECULARIA, Heister, (A. DC.)

551. **S. perfoliata,** A. DC. VENUS'S LOOKING-GLASS.   (**H.** and **C.**)
Borders of dry fields and rocky banks ; frequent. June.-July.

218. CAMPANULA, Tourn.

552. C. RAPUNCULOIDES, L.
Along creeks and gravelly roadsides ; frequent. June.-Aug.
Roadsides on E. Hill, and frequent. North of McLean, on Suy-
der Hill and in Six Mile Cr.

553. *C. rotundifolia, L. HAREBELL.
Cliffs ; frequent. ("Ithaca" in *Sartwell Herb.*) June-Oct.

554. C. aparinoides, Pursh. MARSH-BELLFLOWER. (H. and C.)
Marshes ; frequent. July-Aug.
Indian Spring Marsh. Larch Meadow and elsewhere.

555. C. Americana, L. TALL-BELLFLOWER. (H. and C.)
Shaded banks ; rare. July 15-Aug.
Near the Indian Spring. Bank near S. W. corner of the lake.
(Near Seneca and Crooked Lakes, *Sartwell*; and Conewango Cr.
P. D. K. see *Flora of N. Y.* I, p. 427.)

ERICACEÆ.

219. GAYLUSSACIA, H. B. K.

556. G. resinosa, Torr. and Gray. BLACK-HUCKLEBERRY.
(H. and C.)
Frequent on the dry banks of ravines and high hills. May.
Rarely in marshes ; as South Hill Marsh, Woodwardia Sw. and
Newton's Pond, Junius. Plants are poorly developed and fruit
small in this region.
(*G. dumosa*, and *G. frondosa* in *Sartw. Cat.* 1844, are probably
mistakes.)

220. VACCINIUM, Linn.

557. V. stamineum, L. DEER-BERRY. (H. and C.)
The half-evergreen woods of hills and ravine-banks ; frequent.
May-June.

558. V. Pennsylvanicum, L. DWARF-BLUEBERRY. (H. and C.)
Dry woods or hillsides ; common. May.
Occasional in swampy tracts.

559. V. Pennsylvanicum, L. var. nigrum, Wood, occurs on the North
Pinnacle, Caroline.

560. V. Canadense, Kohm. CANADA-BLUEBERRY.
Sphagnum marshes ; not common. June.
In the marshes of Freeville, McLean and Locke Pond.

561. V. vacillans, Soland. LOW-BLUEBERRY.
Dry woods; frequent. (fruit ripens in Aug. later than *V. Pennsylvanicum*) May.

562. V. corymbosum, L. SWAMP-BLUEBERRY. (H. and C.)
Sphagnum marshes, and swamps, where it is frequent. May 25-
June.

563. V. corymbosum, L. var amœnum, Gray. In Malloryville Marsh.

564. V. corybosum, L. var. atrococcum, Gray, in Larch Meadow and
the Dryden Marshes.

565. **V. Oxycoccus**, L. SMALL CRANBERRY.
  Sphagnum marshes; not common. June–July.
  Freeville, (*Dr. Jordan*)! Malloryville. Round-Marshes. Summit Marsh. Tamarack sw., W. Junius.

566. **V. macrocarpon**, Ait. CRANBERRY. (**H.** and **C.**)
  Sphagnum marshes; frequent. July–Aug.

221. CHIOGENES, Salisb.

567. **C. hispidula**, Torr. and Gray. CREEPING-SNOWBERRY.
                                                    (**H.** and **C.**)
  Cold sphagnum swamps; scarce. May 15–30.
  Brookton Springs. Fir-Tree swamp, S. E. of Danby village. Mud Cr., Freeville. Wyckoff's Swamp, Groton.

222. ARCTOSTAPHYLOS, Adams.

568. **A. Uva-ursi**, Spreng. BEAR-BERRY.
  Banks of ravines; rare. (very rarely flowering here.) May.
  Six Mile Cr. (nearly extinct.) Taughannock, near R. R. where it flowered, May 18, 1885.

223. EPIGÆA, Linn.

569. °**E. repens**, L. TRAILING-ARBUTUS. MAYFLOWER.
                                                    (**H.** and **C.**)
  Gravelly slopes of ravines and moist sandy soil in woods; abundant. Apr. 15.–May 10.
  This is usually abundant in mixed woods of Chestnut and Pine.

224. GAULTIERA, Kalm.

570. °**G. procumbens**, L. WINTERGREEN. CHECKERBERRY. TEA-BERRY.   (**H.** and **C.**)
  Woods and margins of peat-bogs; common. July.

225. ANDROMEDA, Linn.

571. **A. polifolia**, L.
  Sphagnum marshes; infrequent. May 15–30.
  Malloryville marsh. Round-Marshes. Pond Marsh S. E. of Chicago. Tamarack sw. in W. Junius. "Junius," (*Sartwell Herb. and C.*)

572. **A. ligustrina**, Muhl.
  Hill slopes; local and rare. June 15–30.
  South Hill, in the marsh and ¼ mile south of the S. S. 420. Mr. Coville finds it in Chenango Co., and Knieskern gives it as "southern counties" (in Sartwell's Herb.) Our station is probably on its western limits for N. Y.

226. CASSANDRA, Don.

573. **C. calyculata**, Don. LEATHER-LEAF. (**H.** and **C.**)
  Marshes or wet shores; not common. May 10–20.
  Freeville, (*Dr. Jordan*—probably in Woodwardia Sw.) Malloryville. Round-Marsh and Pond Marsh S. E. of Chicago Sta. Dryden Lake valley. Locke Pond. W. Junius Sw.!

227. KALMIA, Linn.

574. **K. latifolia**, L. MOUNTAIN-LAUREL. IVY-BUSH.
Slopes of ravines and evergreen woods ; not common. June.
"Ithaca," (*Bradley, in Sartwell's Herb and C.*; probably at Fall
Creek, near the Fiske-McGraw grounds.) Sparingly in most of the
woods bordering the larger ravines. Formerly in Case. Woods.
On Newfield hills. Somewhat abundant in the great woods on the
west slope of Taft's Hill, Caroline, and about the high cliffs or pin-
nacles at White Church. High hills at W. Danby.

575. **K. angustifolia**, L. SHEEP-LAUREL.
"Ithaca, *Dr. Bradley*," see *Paine's Cat.*, p. 102. "Junius, *Sartwell*," in *Sartwell Herb.* Possibly the Ithaca station was a mistake,
as it has not been observed by any one else.

228. RHODODENDRON, Linn.

576. **R. nudiflorum**, Torr. (*Azalea nudiflora*, L.) PINK-AZALEA.
(H. and C.)
Dry hillside woods ; rarely in swamps ; frequent. May.
The swamp form occurs only at Michigan Hollow, and the Dry-
den-Lansing swamp, where it blooms two or three weeks later than
the other.

577. **R. maximum**, L. BIG LAUREL. RHODODENDRON.
(C. by Dr. Wright.)
Deep swamps ; rare. July.
In the middle of Michigan Hollow Swamp, Danby, over a space
30 meters by 10. Though abundant in Penn., there are only about
a dozen reported stations in N. Y., scattered from the Adirondacks
to Chautauqua Co. The station nearest to ours, a swamp at Italy
Hill, Yates, was detected by Dr. S. H. Wright (not Sartwell, as in
Paine's Cat.)

229. LEDUM, Linn.

578. **L. latifolium**, Ait. LABRADOR TEA. (H. and C.)
Sphagnum marshes ; scarce. June 1-20.
Freeville, (*Dr. Jordan*, probably at Woodwardia Sw.)! Mallory-
ville Marsh. Round-Marshes.

230. CHIMAPHILA, Pursh.

579. **C. umbellata**, Nutt. PRINCE'S PINE. (PIPSISSEWA, of the South.)
(H. and C.)
Dry or moist woods ; common. July 10-30.

580. **C. maculata**, Pursh. SPOTTED PRINCE'S PINE. (H. and C.)
Dry woods ; rare. July.
East shore of Cayuga L., *Ledyard*, 1827, (*in Herb. J. J. Thomas.
"very rare."*) Lansing, (*F. B. Hine.*) West Shore, in Bates
Woods, a few.

231. MONESES, Salisb.

581. **M. uniflora**, Gray. (H and C.)
Deep woods or shaded slopes ; rare. June 15-30.
Near Ludlowville, formerly, (*Mr. Lord*). Knoll near Ellis Hol-
low Sw. under hemlocks, formerly.

232. PYROLA, Tourn.

582. **P. secunda,** L.　　　　　　　　　　　　　　(H. and C.)
Evergreen woods ; abundant. July 1-20.

583. **P. secunda,** L., var. **pumila,** in Paine's Cat., p. 187.
Rare ; in the Fir-Tree swamp at Freeville.

584. **P. chlorantha,** Sw.　　　　　　　　　　　　(H. and C.)
Evergreen woods ; frequent. June 15-July 10.
Case. Woods, Fall Cr., and elsewhere, but only a few in a single
locality.

585. **P. elliptica,** Nutt.　　　　　　　　　　　　(H. and C.)
Woods ; abundant. July 10-30.

586. *P. rotundifolia, L.　FALSE-WINTERGREEN　　(H. and C.)
Woods ; abundant. July 10-30.
Occasionally (as on Thacher's Pinnacle) specimens occur with
oblong leaves and mucronate anthers. (For corrected description
see *Gray's Synopt. Flora.*)

587. **P. rotundifolia,** L., var. **uliginosa,** Gray.
In the Tamarack Sw. S. E. of Savannah, Wayne Co. "Wayne
Co.," (*in Herb. Sartwell.*)

233. PTEROSPORA, Nutt.

588. **P. Andromedea,** Nutt.　PINE-DROPS.　　　　(H. and C.)
Dry slopes of ravines, under Pines ; rare. July 15-Aug. 10.
Coy Glen, 1874. Woods north side of Buttermilk Glen, (*F. H.
Severance,* 1878 ; *F. V. Coville.*) Taughannock ravine, south side,
(*F. C. Curtice,* 1882.) Caroline, slope of North Pinnacle, (*O. E.
Pearce,* 1885)! W. Danby, slope of Thacher's Pinn. "Banks of
Seneca L.," (*Dr. Gray, in Torrey's Flora of N. Y.*)

234. MONOTROPA, Linn.

589. *M. uniflora, L.　INDIAN PIPE.　　　　　　(H. and C.)
Woods ; frequent. July-Aug.

590. **M. Hypopitys,** L.　PINE-SAP.　　　　　　(H. and C.)
Woods ; infrequent. July-Aug.
Fall Creek woods, and very sparingly in most of the old rich
woods of mixed evergreen and hardwood.

PRIMULACEÆ.

235. PRIMULA, Linn.

591. **P. Mistassinica,** Michx.　WILD PRIMROSE.　(H. and C.)
Cliffs, in the crevices or wet moss ; rare. May 15-30.
Fall Creek, on the "Primrose Cliffs," south side of Triphammer
ravine ; where it was detected by Dr. Jordan and the writer after it
had flowered, June, 1871. The high walls of the gray cliffs are pink
with it, when it is in bloom. It grows on the cliffs below, opposite
Rocky Falls (and the wheel-house.) Also south side of Taughannock ravine (detected by Dr. C. Atwood, 1878)! Reported from only
three other principal stations in N. Y. : Along Fish Creek, Annsville,
north-east of Oneida L., where it was known many years ago by
Knieskern and Vasey, and is said to be abundant ; Deep ravine at

the head of Crooked or Keuka L., detected by Dr. Sartwell ; Cliffs at Portage, Genesee R., found by the late Judge Geo. W. Clinton.

236. TRIENTALIS, Linn.

592. * T. Americana, Pursh. STAR-FLOWER. (H. and C.)
Its star-like flowers abundant in most of our cool woods and ravines. May 20-June 10.

237. STEIRONEMA, Raf.

593. * S. ciliatum, Raf. (*Lysimachia ciliata*, L.) (H. and C.)
Low grounds and near streams ; frequent. July–Aug. 15.

238. LYSIMACHIA, Tourn. LOOSESTRIFE.

594. * L. quadrifolia, L. (H. and C.)
Thickets. slopes of ravines, etc.; frequent. June 10-30.

595. L. stricta, Ait. (H. and C.)
Wet places, and on the larger marshes ; frequent. July.

L. PUNCTATA, L., has been growing and slowly extending from the root, for four years on South Hill, near upper Hudson Street. July.

596. L. NUMMULARIA, L.
Established and rather abundant on Lockwood's Flats. In Big Gully. By the creek near Enfield Falls village. One mile north of Enfield Falls. July.

597. L. thyrsiflora, L. (H. and C.)
Marshes in standing water ; infrequent. June.
Near Cayuga St., in Indian Spring Marsh. Danby swamp, on the Sabin farm. Summit marsh. Round-Marshes. Dryden-Lansing swamps. Bear Swamp, Groton. Cayuta L.

ANAGALLIS ARVENSIS, L. (SCARLET PIMPERNEL.) (C. by Dr. Wright.)
was found, 1875, by the R. R. south of Caroline Sta.; not permanent.

239. SAMOLUS, Tourn.

598. S. Valerandi, L., var. Americanus, Gray. WATER PIMPERNEL.
(H. and C.)
Wet shaded places, and dripping rocks ; not common. July.
"Lick Brook? *H. L. Stewart*," 1870, in *Dr. Jordan's Cat.*; same station,(*O. E. Pearce*,)1885. Woods south-east of Sherwood, (*Herb. of Mrs. Prof. Brun.*) 1883, (v. v. 1885)! Limestone falls in brook south of Goodwin's Point. Cayuga and Montezuma Marshes.

OLEACEÆ.

240. LIGUSTRUM, Tourn.

599. * L. VULGARE, L. PRIVET. (H. and C.)
Planted near "Cliff Park." Escaped, and in several places along the top of the cliffs south of Levanna, Cayuga L. shore. June.

241. FRAXINUS, Tourn.

600. * F. Americana, L. WHITE ASH. (H. and C.)
Upland woods ; abundant. May 10-20.

601. F. pubescens, Lam. DOWNY ASH. (H. and C.)
Marshes and alluvial soil ; infrequent. May 10-30.

Indian Spring Marsh; thence to Negundo Woods and Larch Meadow. Cayuga Lake on all the low points; on Frontenac Island, Canoga and Cayuga Marshes. Summit Marsh. From a tree on Sheldrake Point, young shoots from the base of the tree were collected, with pubescent leaves, while leaves from older parts were entirely glabrous, appearing like the next species.

602. **F. viridis**, Michx. GREEN ASH.

Typical specimens from Farley's Point and shore south of Union Springs. Specimens from Lake Ridge, from near Fleming S. H., and from Summit Marsh, show some pubescence.

603. * **F. sambucifolia**, Lam. BLACK ASH.   (H. and C.)

Swamps, or uplands, as by the Campus brook; common. May 1-20.

APOCYNACEÆ.

242. APOCYNUM, Tourn.

604. * **A. androsæmifolium**, L. DOG-BANE.   (H. and C.)

Banks and borders of woods; frequent. July.

605. * **A. cannabinum**, L. INDIAN HEMP.   (H. and C.)

Dry beds of creeks and lake-shores, where it is often prostrate; common. July.

243. VINCA, Linn.

606. * **V. minor**, L. PERIWINKLE. CREEPING-MYRTLE.

Spreading only from the root, near yards and cemeteries; but established in the wildest part of Enfield ravine, below the falls.

ASCLEPIADACEÆ.

244. ASCLEPIAS, Linn.

607. **A. tuberosa**, L. BUTTERFLY-WEED.   (H. and C.)

Dry slopes of ravines and banks, where it is frequent. July.
Case. Cr., Fall Cr., and all the ravines. Abundant on Cayuga L. shore, McKinney's to Aurora.

608. * **A. incarnata**, L. SWAMP-MILKWEED.   (H. and C.)

Low grounds; common. July-Aug.

609. **A. incarnata**, L. var. **pulchra**, Pers., occurs rarely.

610. * **A. Cornuti**, Decaisne. COMMON MILKWEED.   (H. and C.)

By streams and roadsides; common. July.

611. **A. phytolaccoides**, Pursh. WOOD-MILKWEED.   (H. and C.)

In woods; frequent. July.

612. * **A. quadrifolia**, Michx. FOUR-LEAVED MILKWEED.

     (H. and C.)

Ravines and slopes; frequent. June 15-July 15.
A specimen from Franklin's ravine, (*F. H. Parsons*), has all the leaves in pairs 2-3 cm. apart.

GENTIANACEÆ.

245. GENTIANA, Tourn.

613. **G. crinita**, Froel. FRINGED GENTIAN.   (H. and C.)

Wet marshy places; rare. Sept.

"Ithaca" (*in Herb. of Professor J. J. Thomas,*) Case. Cr. (*Dr. Jordan*)! South of Fall Cr., formerly near the present site of the windmill on the Fiske-McGraw grounds. Sherwood, Cayuga Co. (*Miss Howland and Mrs. Brun.*) Lowery's Pond, Junius.

614. **G. quinqueflora,** Lam. FIVE-FLOWERED GENTIAN. (C.)
Wet gravelly banks; abundant in places. Sept.
Fall Cr. Case. Cr., south of Willow Pond Race, and on the Pine Knoll. Very abundant near the Dart Woods, Ball Hill, and elsewhere on the higher hills. Taughannock ravine, south side. Below Sheldrake. Big Gully, and elsewhere.

615. **G. Saponaria,** L. (C.)
Apparently rare. In a small ravine S. E. side of Pony Hollow. "Ledyard" (*Herb. J. J. Thomas.*) Sept.

616. **G. Andrewsii,** Griseb. (H.)
Moist shaded banks, or brooksides; scarce. Sept.
South Hill beyond the "Incline" 1874. Fleming Meadow. Dryden L. valley. Myers Pt. north of Salmon Cr.

617. **G. linearis,** Froel. (*G. Saponaria,* L. var. *linearis,* Gray. (Junius, 1827, (v. s. *in Herb. J. J. Thomas.*)

246. BARTONIA, Muhl.

618. **B. tenella,** Muhl. is in Sartwell Herb. from Junius.

247. MENYANTHES, Tourn.

619. **M. trifoliata,** L. BUCKBEAN.
Sphagnum bogs; infrequent. May 20-30.
Freeville. Malloryville. Round-Marshes. Near the Pond S. E. of Chicago Sta. Summit Marsh. "Junius" (*Sartw. H. and C.*)!

POLEMONIACEÆ.

248. PHLOX, Linn.

PHLOX PANICULATA, L. and P. MACULATA, L. have appeared at Ithaca and McLean as scapes to meadows and roadsides.

620. * **P. divaricata,** L. BLUE PHLOX. (H. and C.)
Rich woods; frequent. May-June.
Fall Cr. Negundo Woods. Ravines of the lakeshore. Woods near Freeville, and elsewhere. A form in Taughannock ravine has elongated leaves, more than usually glandular-pubescent, and white corollas with obovate entire lobes. The ordinary form often has entire corolla lobes,

621. **P. subulata,** L. MOSS-PINK. (C.)
Dry grassy or rocky banks, where it is common. May.
Along all our ravines. Abund. on South Hill south of S. S. 420. Slopes along the lake shore are pink with it in May. It also occurs on the high pinnacles in Danby and Caroline. Rarely occurs with white flowers, (*A. L. Coville.*)

HYDROPHYLLACEÆ.

249. HYDROPHYLLUM, Tourn. WATER-LEAF.

622. * **H. Virginicum,** L. (H. and C.)
Ravines and rich woods; abundant. June.

623. **H. Canadense**, L. (**H.** and **C.**)
Rich woods; frequent. A little later than the preceding.
Six Mile Cr. (Beechwoods.) Negundo Woods. Woods near
Freeville, McLean, Locke Pond, Danby and elsewhere.

BORRAGINACEÆ.

250. CYNOGLOSSUM, Tourn.

624. *C. OFFICINALE, L. HOUNDS-TONGUE. "TORY-WEED,"—a
name local in Cayuga Co. many years ago. (*J. J. Thomas.*)
(**H.** and **C.**
A common roadside and pasture weed. June–July.
A form with pink flowers occurs occasionally. White-flowered
plants grow near Hemlock Brook, Groton; also at "Penn Yan,"
(*Sartwell in Herb.*)

625. **C. Virginicum**, L. WILD COMFREY. (**C.**)
Woods on the higher hills; not common. June-July.
Turkey Hill, eastward to Ringwood. Dart Woods. High hills
of Newfield, Danby and Caroline.

251. ECHINOSPERMUM, Swartz.

626. **E. Virginicum**, Lehm. (*Cynoglossum Morisoni*, DC. BEGGARS-
LICE. (**H.** and **C.**)
Woods and low thickets; common. July–Aug.

627. E. LAPPULA, Lehm. STICKWEED. (**H.** and **C.**)
Sandy or gravelly soil, lake shores and villages; frequent.
July–Sept.
Near the Armory, and on Buffalo St. Near Fall Cr. Mills. Fre-
quent on the sandy points of Cayuga L. and along the R. R.

252. MERTENSIA, Roth.

628. **M. Virginica**, DC. VIRGINIAN COWSLIP. BLUE-BELLS.
Rich alluvial soil; scarce. Apr. 25–May.
Taughannock ravine, (*Dr. Jordan.*)! Negundo Woods, (*Professor
Branner*)! "Valley south of Buttermilk Falls," (*O. L. Taylor*)! Bank
of Negunena Cr., near the Fleming S. H., and near mouth of En-
field Cr. Near Ellis Hollow Swamp, (*F. T. Wilson*)!

253. MYOSOTIS, Linn.

629. M. PALUSTRIS, With. TRUE FORGET-ME-NOT. (**H.** and **C.**)
In water, in marsh north of Fall Cr. and also along Cayuga St.
June 15–July.

630. **M. laxa**, Lehm. (*M. palustris* With. var. *laxa*, Gr.)
In wet places; abundant. June-Sept.
Flowers occasionally pink.

631. M. ARVENSIS, Hoff. SCORPION-GRASS. (**H.** and **C.**)
Shaded sandy soil; scarce and probably introduced. June.
Cemetery, Ithaca. (*F. L. Kilborne.*) Bushy Point, Cayuga L.

632. M. COLLINA, Hoff. (apparently this species) found by Mr. Kilborne in 1882; appears each year.

633. **M. verna**, Nutt.
Dry, barren woods near the lake; occasional. June.

Farley's Pt., (*Herb. J. J. Thomas.*) From Renwick Farm to Willet's Sta. Near Taughannock and Trumansburg ravines. In the bottom of Franklin's ravine it grows 30 cm. in height and still taller in the rich wooded slope, east side of the valley at Watkins.

254. LITHOSPERMUM, Tourn.

634. L. ARVENSE, L. CORN-GROMWELL. (H. and C.)
A common weed in grain fields, etc., in dry soil. May, June.

635. L. OFFICINALE, L. COMMON-GROMWELL.
Roadsides and meadows ; not common. June-July.
Near Ludlowville, N. Y., (*Mr. Lord.*) Aurora to Chapel-Corners. especially near the latter. Near Wood's Sta. in Scipio.

636. L. latifolium, Michx. (H. and C.)
Shaded ground near streams ; rare. June.
Salmon Cr. ravine, north bank with *Ionidium*, 1871 to 1885.
Fall Cr. on island above Forest Home, (*F. B. Hine,*) 1875.

[L. canescens, Lehm. Near Auburn, N. Y. *I. H. Hall, in Paine's Cat.,* p. 114.]

255. ONOSMODIUM, Michx.

637. O. Carolinianum, DC.
Hillsides ; local. July.
South Hill from near the reservoir to the Quarry Woods beyond the " Incline ;" frequent over this area.

[O. Virginianum, DC. "Seneca L.," *Sartwell's Cat. of* 1884.]

256. SYMPHYTUM, Tourn.

638. *S. OFFICINALE, L. COMFREY. (H. and C.)
Rich soil along streams ; not uncommon. June-July.
Casc. Cr., abund. above Judd's Falls. Six Mile Cr. Neguæna valley to Newfield. Myers Point and near Lake Ridge Sta. abundant.
Fall Cr., E. of Freeville and elsewhere.

BORRAGO OFFICINALIS, L., appeared on the Campus in 1878 and in 1880.

ASPERUGO PROCUMBENS, L., was found in 1885 near E. Hill S. H. by *F. V. Coville.*

CONVOLVULACEÆ.

IPOMEA, Linn.

I. COCCINEA, L., was growing, in 1885, in a large mass by the old road crossing the marsh to the steamboat landing.

I. PURPUREA, Lam. MORNING-GLORY.
Springing up in the same place for several years, perhaps from self-sown seeds. July-Aug.
Six Mile Creek near Cayuga St. bridge. North of Ithaca, by R. R. near the Glass-works, and by an old road. U. Springs near the mill, and the R. R.

I. NIL, Roth., occasionally seen with the preceding.

257. CONVOLVULUS, Linn.

639. C. ARVENSIS, L. BIND-WEED. (H. and C.)
Roadsides and old fields ; scarce. June–July.

Fall Cr., on the Circus Common. Hazen St., Ithaca, (*F. V. Coville.*)

640. * **C. spithamæus**, L. (*Calystegia spithamæa*, Pursh.) HEDGE-
BIND-WEED.　　　　　　　　　　　　　　　　(H. and C.)
Declivities of ravines and hills; not common. June.
Fall Cr., lower part of ravine, and bank toward the " Nook."
Enfield ravine, north and south bank. W. Danby Pinnacles.
Taughannock. Lake shore, north of King's Ferry.

641. * **C. sepium**, L. (*Calystegia sepium*, R. Br.)　　(H. and C.)
Banks of streams and lakes; common. June–Sept.
A form (possibly *var. repens*, Gray.) on Montezuma marshes, has
stems scarcely twining, leaves narrowly hastate, pubescent.

258. CUSCUTA, Tourn. DODDER.

642. **C. tenuiflora**, Englm.
Shores or low grounds; rare. Aug.–Sept.
Union Springs (on *Mentha piperita*, L.) Montezuma, on mead-
ows north-east of village, abundant (on *Aster paniculatus*, Lam.)

643. **C. inflexa**, Englm.
Hillsides and thickets; scarce. Aug., Sept.
Fall Cr. (on *Asters* and *Solidagos*.) Hillside near White Church
(on *Daisies* and *Ceanothus*.) W. Danby, by the hill road near
Thacher's Pinnacle.

644. **C. Gronovii**, Willd.　　　　　　　　　　　　　(H. and C.)
Near marshes; frequent (on various plants.) Aug.

645. C. EPILINUM, Weihe. Frequent on the cultivated flax on farms
E. of Owasco L. (*Mr. Kilborne*); also rarely appears in flax fields
between Cayuga L. and Seneca L. This is the injurious "FLAX-
DODDER " of Eur.

SOLANACEÆ.

259. SOLANUM, Tourn.

646. S. NIGRUM, L. COMMON NIGHTSHADE.　　　　(H. and C.)
Shady or waste ground; not common. July–Sept.
Casc. Cr., (*H. L. Stewart*)! Fall Cr. and elsewhere about Ithaca
and Cayuga L. shore.

647. S. DULCAMARA, L. NIGHTSDADE BITTER-SWEET.　(H. and C.)
Hedges, thickets and swamps; common. June–Aug.

260. PHYSALIS, Linn.

648. **P. Virginiana**, Mill. (*P. viscosa*, of Man., p. 382.) GROUND
CHERRY.　　　　　　　　　　　　　　　　　(H and C.)
Dry banks and fields; occasional. July–Sept.
Fall Creek, in the ravine; in the field south of Ithaca Fall, and
elsewhere.

[P. pubescens, L. (*P. obscura*, Mx.) at Penn Yan, (*Sartwell, Herb.
and C.*)]

261. LYCIUM, Linn.

649. * L. VULGARE, Dunal. MATRIMONY-VINE.　　　(H. and C.)
Fall Creek, on the Circus Common. Near Freeville. June.

262. DATURA, Linn.

650. * D. STRAMONIUM, L. STRAMONIUM. (H. and C.)
Waste grounds, and lake-shore ; frequent. Aug.-Sept.

651. D. TATULA, L., with the preceding. PURPLE STRAMONIUM.
(H. and C.)

LYCOPERSICUM ESCULENTUM, Mill., (TOMATO.) is occasional on the
lake shore and near Ithaca.

[NICANDRA PHYSALOIDES, Gærtn., at Penn Yan, (*Sartwell, Herb.
and C.*)]

SCROPHULARIACEÆ.

263. VERBASCUM, Linn.

652. * V. THAPSUS, L. MULLEIN. (H. and C.)
Fields ; common, especially in old fields along the lake and the
ravines. July-Sept.

653. V. BLATTARIA, L. MOTH-MULLEIN. (H. and C.)
Fields and pastures ; frequent. June-Aug.
Form with yellow flowers near Six Mile Cr., White Church
and Danby; form with white, Six Mile Cr., McLean and elsewhere.

264. LINARIA, Tourn.

654. * L. VULGARIS, Mill. TOAD-FLAX. (H. and C.)
Gravelly banks and fields ; frequent. June-Oct.

655. L. ELATINE, Mill. (C.)
Shores, etc.; rare. June-Sept.
"Shores of Cayuga L." (probably at Sheldrake), (*Gray in Torrey's Flora.*) Sheldrake Pt., (*Mr. Lord.*)! Farley's Pt., several
places, and by the R. R., just east. (Streets of Geneva, *Sartwell.*)

265. COLLINSIA, Nutt.

656. C. verna, Nutt.
"Ithaca, *Aikin,*" (*Torrey's Flora of N. Y.*) Although it is certain that Aikin visited Cayuga L. about fifty years since, no specimen of *Collinsia* from Ithaca exists in Dr. Torrey's herbarium, and
this beautiful flower has never been rediscovered here.

266. SCROPHULARIA, Tourn.

657. S. nodosa, L. FIGWORT. (H. and C.)
Rich soil along streams, and in ravines ; frequent. June-Aug.

267. CHELONE, Linn.

658. * C. glabra, L. TURTLE-HEAD. (H. and C.)
Marshes, meadows and swamps ; frequent. Aug.-Sept.

268. PENTSTEMON, Mitchell.

659. * P. pubescens, Soland. BEARD-TONGUE. (H. and C.)
Rocky places ; abundant along ravines and cliffs. June-July.
Near mouth of Fall Cr. ravine, (*Pursh.*, 1807.)! One of the few
plants of this region mentioned by him in his Journal of Travels.
It is particularly abundant and beautiful on the ledges of the lake
shore cliffs, taking the place of the earlier Columbines.

269. MIMULUS, Linn.

660. *M. ringens, L. MONKEY-FLOWER. (H. and C.)
Marshes and brooksides ; common. July, Aug.

270. GRATIOLA, Linn.

661. G. Virginiana, L. (H. and C.)
Muddy shores and fields ; frequent. July-Aug.

271. ILYSANTHES, Raf.

662. I. gratioloides, Benth. FALSE PIMPERNEL. (H. and C.)
Wet shores ; scarce. July-Aug.
Lake marsh, (Dr. Jordan.) Shore W. of Inlet. Mouth of Salmon
Cr. Near R. R., east of Judd's Falls. Cortland Marl Ponds.

272. VERONICA, Linn.

663. V. Virginica, L. CULVER'S-ROOT.
Sheltered banks ; rare. July 15-Aug.
Near Fall Cr, north of Ithaca, (Dr. Underwood, of Groton.)
Woods near end of Farley's Point. Utt's Point. The wild bank
beyond the Fleming S. H. "Junius," (Sartæ. Herb. and C.)

664. * V. Anagallis, L. WATER-SPEEDWELL. (H. and C.)
Usually in water ; rare or local. Aug.
Eddy Pond, 1882. Marl Ponds of Cortland. Abundant in outlet
of Lay's Iron Spring, west of Black Lake.

665. * V. Americana, Schwein. BROOK-LIME. (H. and C.)
In rivulets and ravines ; frequent. June-July.

666. * V. scutellata, L. MARSH-SPEEDWELL. (H. and C.)
Marshy places ; frequent. June-Oct.

667. * V. officinalis, L. COMMON SPEEDWELL. (H. and C.)
Roadsides, old fields and pastures ; frequent. June-July.
Particularly abund. on West Hill, Turkey Hill and near McLean.

668. V. serpyllifolia, L. THYME-LEAVED SPEEDWELL. (H. and C.)
Fields and lawns ; common. May-Oct.

669. V. peregrina, L. (H. and C.)
Cult. ground ; common. May-June.

670. V. ARVENSIS, L. CORN-SPEEDWELL. (H. and C.)
Cult. grounds ; common. (Gray, in Synopt. Flora, considers it
" rather rare." May-June.

671. V. BUXBAUMII, Tenore.
Cult. ground ; local, but persistently spreading. Mar.-April.
W. Hill in Sanford's garden, (Prof. J. H. Comstock, 1882)!

[Buchnera Americana, L., in Gorham, Ont. Co., (Sartwell, in Herb.
and C.)]

273. GERARDIA, Linn.

672. G. Pedicularia, L. (H. and C.)
Dry woods near most of our ravines ; frequent. Aug.-Sept.
Particularly abundant, certain years, in Cornell's woods near
Case. Cr. Beside the ravine stations it occurs in Ball Hill woods,
and Thacher's Pinnacle, in Danby ; also the declivities near White
Church.

673. **G. flava**, L. DOWNY FALSE-FOXGLOVE. (**H.** and **C.**
Open dry woods of Oak ; frequent. July-Aug.

674. **G. quercifolia**, Pursh. SMOOTH FALSE-FOXGLOVE. (**H.** and **C.**)
The larger open woods along ravines and on hills ; not rare.
July-Aug.
Coy Glen. Bates Woods west of Cayuga L. Hills of Danby and
Caroline, and elsewhere.

675. **G. purpurea**, L. PURPLE GERARDIA. (**C.**)
Shores ; rare. Aug.-Sept.
Farley's Point, rather abundant along the south shore. Marsh
about Newton's Ponds, Junius.

676. **G. tenuifolia**, Vahl. SLENDER GERARDIA. (**H.** and **C.**)
Sandy banks ; rare. Aug., Sept.
North bank of Salmon Cr. below Ludlowville.

274. CASTILLEIA, Mutis.

677. * **C. coccinea**, Spreng. PAINTED-CUP. (**H.** and **C.**)
" Near Aurora," (*Dr. Alex. Thompson, Cat. of* 1842.) Not now
known within limits. Miss Arnold sends specimens (some scarlet,
some yellow,) from Painted Post.

275. PEDICULARIS, Tourn.

678. * **P. Canadensis**, L. LOUSEWORT. (**H.** and **C.**)
Open woods, and ravines ; common. May.

679. **P. lanceolata**, Michx.
Banks and marshy places ; rare. Aug. 20-Sept. 20.
Small marsh west of Pleasant Grove Cemetery. East of the
Clock Factory, Ithaca. Ludlowville, near the R. R.north of Sal_
mon Cr.

276. MELAMPYRUM, Tourn.

680. **M Americanum**, Michx. COW-WHEAT. (**H.** and **C.**)
Dry woods, near ravines and on hill ; common. June-July.

OROBANCHACEÆ.

277. APHYLLON, Mitchell.

681. **A. uniflorum**, Gray. BROOM-RAPE. (**C.**)
Moist shaded places ; scarce. May 20-June.
Fall Cr. near Beebe Pond. Case. Woods. Turkey Hill, in woods.

278. CONOPHOLIS, Wallr.

682. * **C. Americana**, Wallr. CANCER-ROOT. (**H.** and **C.**)
Woodlands on the hills ; not rare. July.
" Ledyard, in woods of Red and Yellow Oak; rare," (in *Herb. J.
J. Thomas.*) Turkey Hill, (*F. H. Severance,*) 1878 ! Woods north
of Forest Home, (*F. C. Curtice.*) High hills in Caroline. Danby,
hills east of village. Woods in Ovid.

279. EPIPHEGUS, Nutt.

683. * **E. Virginiana**, Bart. BEECH-DROPS. (**H.** and **C.**)
Beech woods ; frequent. Sept.-Oct.

LENTIBULACEÆ.

280. UTRICULARIA, Linn. BLADDERWORT.

684. U. vulgaris, L. COMMON BLADDERWORT. (H. and C.)
Aquatic ; frequent. July 15-Aug.

Still water bayous near mouth of Fall Cr. and near C. S. R. R. at Glass-works and the Corner-of-the-lake. Pools between the R. R. and the cliffs north of Ludlowville, and north of King's Ferry. Marsh north of Union Springs. "Union Springs," (*in Herb. J. J. Thomas.*) Black Lake and Cayuga Marshes.

685. U. minor, L.
Specimens with foliage and bladders of this species, but flowers not seen, occur in Summit Marsh and Black Lake. Flowering time for this latitude is, June 10 July 15.

686. U. intermedia, Hayne.
Marshes and shallow water; scarce. July 15-Aug. 15.
Swamp north of Dryden L. Summit Marsh. Black Lake.

687. U. cornuta, Michx. ; rare, on muddy shores or marshes, July, is in *Sartwell, Herb. and C.* from "Junius." It occurs just north of the Junius Pout Pond in abundance.

281. PINGUICULA, Tourn.

688. P. vulgaris, L. PINGUICULA, BUTTERWORT.
Wet cliffs, in ravines ; rare. June 10-30.

Cascadilla Creek, (*Dr. Jordan.* 1869) ! where it has continued on both sides of the Glen Pond ; also grows farther down the ravine. Fall Creek on the "Primrose Cliffs;" also opposite Rocky Falls. Taughannock ravine, south side with *Saxifraga aizoides*, L. The only other stations reported for N. Y. are on the Genessee River—formerly "in meadows below the Falls at Rochester," (*Dewey, Carey* and others,) and still existing below Portage Falls, (*Dewey ;* and "*Plants of Buffalo,*" 1882.)

BIGNONIACEÆ.

282. CATALPA, Scop., Walt.

689. * C. BIGNONIOIDES, Walt. (CATALPA) is growing spontaneously under the cliff west of the Ithaca Cemetery.

ACANTHACEÆ.

283. DIANTHERA, Linn.

690. D. Americana, L. WATER-WILLOW.
Shores, and borders of marshes; scarce. July-Aug.
Farley's Point. From Cayuga Bridge to Montezuma, along Seneca R.

VERBENACEÆ.

284. PHRYMA, Linn.

691. P. Leptostachya, L. LOP-SEED. (H. and C.)
Woods and thickets ; frequent. July.

285. VERBENA, Tourn.

692. * V. urticæfolia, L. WHITE VERVAIN. (H. and C.)
Low grounds ; common. July–Sept.
Form with pink flowers, at Mud Cr. Form (hybrid with *V. has-tata*, L. ?) in Trumansburg ravine, with flowers lavendar or pink. (For discussion of hybrid Verbenas, see Engleman in *Am. Jour. Sci.*, XLVI.) Hybrids with purple and with blue flowers, are from Penn Yan, in *Herb. Sartwell*.

693. V. hastata, L. BLUE VERVAIN. (H. and C.)
Low grounds ; common. July–Sept.

LABIATÆ.

TRICHOSTEMA DICHOTOMUM, L., (BLUE CURLS,) was found by F. B. Hine, in 1879, on the high bank north of Beebe Pond. Not seen since.

286. TEUCRIUM, Linn.

694. * T. Canadense, L. GERMANDER. (H. and C.)
Along the lake shores and marshes ; frequent. July–Sept.
Form with white flowers near Seneca L., (*Sartwell, H. and C.*)

287. COLLINSONIA, Linn.

695. * C. Canadensis, L. HORSE BALM. H. and C).
Woods and moist shaded places ; common. July–Aug.

288. MENTHA, Tourn.

696. * M. VIRIDIS, L. SPEARMINT. (H. and C.)
Roadsides and wet places ; common. Aug.-Sept.

697. * M. PIPERITA, L. PEPPERMINT. (H. and C.)
Dry beds of streams and elsewhere ; common.

698. M. PIPERITA, L. var. SUBHIRSUTA, Benth. ; apparently this form occurs among the willows, south end of Cayuga L. east of Fall Cr.

699. * M. Canadensis, L. WILD-MINT. (H. and C.)
Along water courses and marshes ; common. July–Sept.

289. LYCOPUS, Tourn.

700. L. Virginicus, L. BUGLE-WEED. (H. and C.)
Wet places, often in ravines ; common. Aug., Sept.
Specimens growing at the foot of the cliffs north of Ludlowville Sta. exhibit well formed tubers at the ends of numerous slender drooping branches.

701. L. sinuatus, Ell. (*L. Europaeus*, L., var. *sinuatus*, see Man., p. 345.) (H. and C.)
Shores and marshes ; common. July–Sept.
One of the commonest plants on Cayuga and Montezuma Marshes.

[HYSSOPUS OFFICINALIS, L. Frequent in Yates Co., *Sartwell, H. and C.*]

290. PYCNANTHEMUM, Michx. MOUNTAIN-MINT.

702. P. lanceolatum, Pursh.
Borders of West Junius marshes, about the ponds. Aug. 15–Sept.

703. **P. incanum**, Michx. (**H.** and **C.**)
Dry woods near the lake ; not common. July–Aug.
Near Pleasant Grove Brook. Woods and slopes from McKinney's
north. S. W. corner of Cayuga L. North of Ludlowville, and of
King's Ferry. Paine's Creek.

291. THYMUS, Tourn.

704. T. SERPYLLUM, L. CREEPING THYME. WILD THYME.
Beside the road between W. Groton and Locke ; well established,
(*F. L. Kilborne,*) Aug. 15, 1881.

T. VULGARIS, L. Found one year, (1882,) growing spontaneously
on the Fiske-McGraw grounds, (by *F. L. K.*)

292. CALAMINTHA, Tourn.

705. C. ACINOS, Clairv.
Along the road east of Ithaca Cemetery. (*Mr. Lord*, 1881) ! It
also occurs in the rocky part of the cemetery.

706. **C. Clinopodium**, L. BASIL. (**H.** and **C.**)
Fields and pastures ; common. July.

293. MELISSA, Tourn.

707. *M. OFFICINALIS, L. COMMON BALM. (**C.**)
Roadsides and banks ; scarce. July–Aug.
By road "near Newfield," 1878, (*Trelease* and *Severance.*) By
R. R. near King's Ferry. Big Gully, below crossing of second road.

294. HEDEOMA, Pers.

708. *H. pulegeoides, Pers. PENNYROYAL. (**H.** and **C.**)
Dry ground ; abundant. Aug.

295. MONARDA, Linn.

709. *M. didyma, L. BEE-BALM. SCARLET BALM. (**H.** and **C.**)
Rich woods, borders of low grounds ; frequent. July–Sept.
Lake Marshes, 1873, (*Kellermann*) ! Negundo Woods. Case. Cr.
toward Ellis Hollow, through Ringwood, to the Freeville and
Round-Marsh region. White Church valley. Danby. W. Danby
to Signer's Woods. Taughannock and other ravines.

710. M. Clinopodia, L. (*M. fistulosa,* L., in part, of the Man. *Pycnanthemum Monardella*, Michx., Flora, II, 8, t. 34.)
Shaded ravines ; rare. July–Aug.
Six Mile Cr., below Green-Tree Falls. Trumansburg ravine.
This plant is characterized by its taller, slender, nearly smooth or
somewhat villous stems ; thin, slender-petioled, coarsely-serrate
leaves ; throat of calyx sparingly hirsute and corolla dull-white or
pink, and rather shorter than in *M. fistulosa L.* (See *Gray, Synopt.
Flora*, II, (1), p. 374.)

711. *M. fistulosa, L., var. mollis, Benth., (see *Gray, Synopt. Flora*,
II, (1), p. 374.) WILD BERGAMOT. (**H.** and **C.**)
River-banks and shores ; frequent.
Negundo Woods, and up the valley to Newfield. Renwick.
Myers Point. Paine's Cr. Union Springs. Taughannock, etc.

712. **M. fistulosa**, L... var. **rubra**, Gray? (or possibly a hybrid, as Dr. Gray suggests.) On the wild bank beyond the Fleming S. H., several variable plants with large heads and rose-colored corollas, suggesting *M. didyma* in habit.

296. BLEPHILIA, Raf.

713. **B. hirsuta**, Benth. (C. by Dr. Wright.)
Shaded ravines and banks of streams ; scarce. July.

Fall Cr., Beebe Pond, on the island. Six Mile Cr., in the amphitheatre above the " Narrows." Enfield in ravine below the Falls. Near the south shore of Cayuga L.

297. LOPHANTHUS, Benth. GIANT HYSSOP.

714. **L. nepetoides**, Benth. (C.)
Rich soil and shaded banks ; scarce. Aug.
S. W. corner of the Lake. Negundo Woods. South Hill.

715. **L. scrophulariæfolius**, Benth. (H. and C.)
With the preceding ; scarce. Aug.
By road toward Renwick. Near S. W. corner of the lake. Paine's Cr.

298. NEPETA, Linn.

716. *N. CATARIA, L. CATNIP. (H. and C.)
Waste places and shores ; common. July-Oct.

717. N. GLECHOMA, Benth. GROUND IVY. GILL. (H. and C.)
Waste places and in ravines ; frequent. Apr.-Aug.

299. SCUTELLARIA, Linn. SKULL-CAP.

718. * S. lateriflora. L. (H. and C.)
Low grounds ; frequent. Aug.

719. *S. galericulata, L. (H. and C.)
Borders of marshes ; frequent. July, Aug.
An erect form often grows freely on the R. R. embankments near the lake.

300. BRUNELLA, Tourn.

720. * B. vulgaris, L. BRUNELLA. (H. and C.)
Fields ; common. July-Oct.
Forms with white flowers near Mud Cr. and Beaver Cr.; with pink flowers near Beaver Cr.

PHYSOSTEGIA VIRGINIANA, Benth., formerly grew by the brook in the fields east of the R. Catholic cemetery near Ithaca. It continued from 1873 till 1884, but seems to be now extinct. Probably introduced.

301. MARRUBIUM, Tourn.

721. *M. VULGARE, L. HOREHOUND. (H. and C.)
Reported from near Enfield Falls (*Severance* and *Trelease*) ; and " near Ithaca," (*Mr. Lord.*) Abundant in a field S. W. of Paine's Cr., also occurs in the ravine. July-Aug.

302. LEONURUS, Linn.

722. * L. CARDIACA, L. MOTHERWORT.
Fields and waste places ; common. July-Sept.

303. LAMIUM, Tourn.   DEAD-NETTLE.

723. L. AMPLEXICAULE, L.

     Fields ; scarce. Apr.-Nov.

     E. Hill, on the campus and farm. Fall Cr. Circus Common. Near Big Gully.

L. MACULATUM, L., has escaped by the road north of Forest Home.

304. GALEOPSIS, Linn.

724. *G. TETRAHIT, L.   HEMP-NETTLE.   (H. and C.)

     Waste grounds ; not common. July-Sept.

     Fall Cr. below the mills. Freeville. Near Groton. Locke Pond.

305. STACHYS, Tourn.

725. *S. aspera, Michx., (*S. palustris*, L., var. *aspera*, Gray.)

                                                (H. and C.)

     Abundant on both shores of Cayuga L. July-Sept.

726. S. aspera, Mich., var. glabra, Gray.

     In herb. of Prof'r J. J. Thomas from Cayuga L. Lake marsh near Ithaca. Near McKinney's.

PLANTAGINACEÆ.

306. PLANTAGO, L.   PLANTAIN.

727. P. cordata, Lam., is in herb. of Professor Thomas, collected in Ledyard, 1827.

728. *P. MAJOR, L.   (H and C.)

     Fields, shores and pastures ; frequent. June-Sept.

     This species is distinguished from the following by its thicker, shorter spike, its short, ovoid, obtuse capsules and ovate, obtuse, scarious margined sepals, and its leaves often dull green and hairy. It is less common than the next species in our Flora, and has long been confused with it everywhere. The var. *minima*, (*P. minima.* DC.) occurs in cold, dry soils near Freeville, Summit Marsh and on the higher hills.

729. P. Rugelii, Decaisne. (*P. Kamtschatica*, of Man., p. 311,—not of Cham.)

     The commonest roadside and yard plantain. June-Sept.

     The spikes are longer than in *P. Major*, L., less dense, capsules oblong-cylindrical and sepals oblong and acute. The leaves are large, usually shining, bright green, the petioles often tinged with purple. Dr. Gray regards *P. Rugelii*, Decaisne, as " truly indigenous, as no trace of it is found in the Old World." The spike of this species is frequently fasciated at the summit.

730. *P. LANCEOLATA, L.   RIBWORT.   (H. and C.)

     Fields ; common. June-Aug.

731. P. MEDIA, L.

     Grounds of the Pres.-White place since 1883 (*F. L. Kilborne*) ! Specimens were transferred to the Sage Building court. It is recognized by its broadly oval, hairy leaves lying flat upon the turf, and by its conspicuous, usually pale-purple stamens.

ILLICEBRACEÆ.

307. ANYCHIA, Michx.

732. **A.** dichotoma, Michx. (C.)

Dry banks and cliffs, where it is frequent. July.

Fall Cr. and Casc. Cr. Near McKinneys. Salmon Cr. Taughannock ravine. High cliffs at King's Ferry and elsewhere.

308. SCLERANTHUS, Linn.

733. S. ANNUUS, L.

Field south of Coy Glen. Sheldrake Point. June–July.

AMARANTACEÆ.

309. AMARANTUS, Tourn.

734. A. PANICULATUS, L. RED AMARANTH.

Roadsides. Near Renwick and on the Univ. campus. July.

735. A. RETROFLEXUS, L. PIGWEED. (H.)

Fields; everywhere common. July–Sept.

736. A. CHLOROSTACHYS, Willd.

Less common than the preceding, but frequent.

737. A. ALBUS, L. (H. and C.)

Fields and shores; abundant.

738. A. BLITOIDES, Watson. TUMBLE-WEED.

An immigrant from the far west. July–Aug.

Union Springs, by the R. R., near the steamboat landing, 1881. West of Cayuga Bridge, 1885.

"Prostrate or decumbent, the slender stems becoming a foot or two long, glabrous or nearly so; leaves broadly spatulate to narrowly oblanceolate, attenuate to a slender petiole, an inch long or usually less; flowers in small contracted axillary spikelets; bracts nearly equal, ovate-oblong, shortly acuminate, 1 to 1½ lines long, little exceeding the oblong obtuse and mucronulate or acute sepals; utricle not rugose slightly longer than the sepals; seed nearly a line broad." (*Watson, Proc. Amer. Acad.,* XII, p. 273). Somewhat resembles *A. albus,* L. On the western plains it becomes one of the "tumble-weeds," in autumn, being broken off at the root, and driven before the wind.

CHENOPODIACEÆ.

310. CHENOPODIUM, Linn.

739. C. ALBUM, L. PIGWEED. (H. and C.)

Fields; very common. July–Sept.

740. C. URBICUM, L. (H.)

Recently introduced; Ithaca, Cayuga St., 1880; Aurora St., 1882.

741. C. HYBRIDUM, L. (H. and C.)

Waste places, and the lake shore; frequent. Aug.–Sept.

Appearing as if indigenous on the rocks at Esty Glen, King's Ferry, etc.

742. C. BOTRYS, L. JERUSALEM OAK. (H. and C.)

Dry soil; infrequent. July–Oct.

Fall Cr. mills. Near Inlet by D. L. and W. R. R. U. Springs and Lockwood's Flats, (*Herb. J. J. Thomas.*) Kidder's Ferry. Sheldrake Pt.

311. BLITUM, Tourn.

743. **B. rubrum**, Reich. (*B. maritimum, of Man.*, p. 408.)
Brackish soil ; rare, on meadows N. E. of Montezuma village. Aug.-Sept.

744. **B. capitatum**, L. STRAWBERRY BLITE. (**H.** and **C.**)
Woods and newly cleared soil ; scarce. Aug., Sept.
Near Freeville sphagnum bog. Pasture near the Marl Ponds. Hill east of Pony Hollow. Near Ludlowville, (*Mr. Lord.*) Ledyard, 1827, (*Herb. J. J. Thomas.*)

312. ATRIPLEX, Tourn.

745. **A. patula**, L., var. **hastata**, Gray. Aug.-Oct.
Streets of Ithaca. Sheldrake Pt. Union Springs and Cayuga Bridge. Near the Indian Salt-Spring west of Cayuga marshes, near the " Deer-lick," and near brackish meadows at Montezuma, the stems are bright red in autumn.

745³. **A. patula**, L., var. **littoralis**, Gray.
Buffalo St., Ithaca, and on the marsh; abundant. Union Springs on Farley's Point and Howland's Point.

PHYTOLACCACEÆ.

313. PHYTOLACCA, Tourn.

746. **P. decandra**, L. POKEBERRY. (**H.** and **C.**)
Dry hillsides and pastures ; frequent. Aug.

POLYGONACEÆ.

314. RUMEX, Linn.

747. R. PATIENTIA, L. PATIENCE DOCK.
Ithaca, near Six Mile Cr. bridge, (Cayuga St.) Near W. Junius. June-July.

748. **R. Britannica**, L. (*R. orbiculatus*, Gray's Man. (**H.** and **C.**)
Large marshes, where it is abundant. Aug. Sept.
Ithaca, marsh near the Lake. Summit Marsh. Round Marshes. Dryden L. "Cayuga marshes," (*G. W. C.*, 1864, 18th N. Y. Rep., p. 201;) Very conspicuous on these and Montezuma Marshes in autumn from its large plumes of pink fruit.

749. **R. verticillatus**, L. SWAMP DOCK. (**H.** and **C.**)
Marshes ; abundant. June-July.
Marshes at the head of Cayuga L. Cayuga Marshes, Summit Marsh, and elsewhere.

750. * R. CRISPUS, L. YELLOW DOCK. (**H.** and **C.**)
Fields ; common. June-Sept.

751. R. OBTUSIFOLIUS, L. BITTER DOCK. (**H.** and **C.**)
Roadsides and woods ; frequent. July, Sept.
Occasionally specimens possess red-veined radical leaves.

752. **R. crispus,** L. ✕ **R. obtusifolius,** L.

Specimens with the narrow leaf, nearly that of *R. crispus,* and the valves toothed somewhat less than *R. obtusifolius,* occur near the outlet of Dryden L., near Renwick, and by roadside in Danby.

753. R. CONGLOMERATUS, Murr.

Near the Marl Ponds, Cortland.

754. R. ACETOSELLA, L.   SHEEP SORREL.   (H. and C.)

Fields everywhere.   May–Sept.

[R. ACETOSA, L., at Penn Yan, (*Sartwell in Herb. and Wright in C.*)]

315.   POLYGONUM. Linn.

755. P. ORIENTALE, L.   PRINCES FEATHER.   (H. and C.)

Waste ground and shorés ; occasional.   Aug.

Near Six Mile Cr., south of Ithaca ; near the Inlet.   Cayuga Lake shore, especially south of Cayuga Bridge.

756. **P. Pennsylvanicum,** L.   (H. and C.)

In moist, rich soil ; common.   Aug.-Oct.

757. **P. incarnatum,** Ell.

Marshes and shores ; not common.   July–Sept.

Inlet marshes, (*Dr. Jordan.*) !   Frequent near the foot of Cayuga Lake and along its outlet.   Ledyard 1827, (*Herb. J. J. Thomas.*)

758. P. NODOSUM, Pers.

Scarce near the shores of Cayuga Lake (where it seems indigenous,) at the south end, on Myers Pt.   W. of Cayuga Bridge and on Cayuga Marshes.   Aug.-Sept.

"Stem annual, stout, (1°-4°), mostly glabrous, often sparingly and minutely glandular on the peduncles.   Sheaths not fringed or hairy ; style 2-cleft ; stamens 6 ; leaves narrowly lanceolate, attenuate upwards from near the base and acuminate, cuneate at the base and short petioled, somewhat scabrous with short prickly hairs near the midribs and margins ; spikes oblong and erect, or often linear and nodding, 1′ long or more ; flowers 1″ long, white or rose : akenes ovate less than a line broad."   Chiefly Rocky Mts. and Pacific Coast.   (*Bot. of Cal.,* II., p. 13.)

Mr. Watson writes me that it has been found along railroads in Mass. sparingly.

759. **P. lapathifolium,** Ait., var. **incanum,** Koch.

"Borders of Cayuga L." (*"Chickering and Brewer"* in Man., p. 416.)   Not yet rediscovered here.

760. P. PERSICARIA, L.   (H. and C.)

Common everywhere.   July-Oct.

A very smooth form prostrate and rooting occurs in exsiccated places.   A much branched hairy form with interrupted spikes, flowers and bracts of a brighter pink than usual, grows along Canoga Marshes.

761. **P. Hydropiper,** L.   SMARTWEED.   (C.)

Fields and low grounds ; common.   Aug.-Sept.

762. **P. acre.** H. B. K. (**C.** by Dr. Wright.)
Marshes and shores; frequent. Aug.-Oct.
Specimens on the Inlet Marshes, often with slenderer leaves than in the type.

763. **P. hydropiperoides,** Michx. WATER-PEPPER. (**H.** and **C.**)
Marshes and muddy banks; infrequent. Aug.-Oct.
Marshes near head of the lake, (by Cayuga St.) Along Fall Cr. between Forest Home and Etna. Summit Marsh. Our specimens have the "oblong" leaves so far as noticed.

764. **P. amphibium,** L. (the typical form of Gray's Man.,)
(**H.** and **C.**)
Aquatic; not common. July-Sept.
Dryden L. Marl Ponds. Summit Marsh. Danby, in Jenning's Pond. Cayuga and Montezuma Marshes.

765. **P. Muhlenbergii,** Watson. (*Proc. of Amer. Acad.,* XIV, p. 295, 1879.) (**C.** by Dr. Wright.)
Gravelly banks and shores; frequent. Sept.-Oct.
By the old R. R. track near the light-house and shore at the head of Cayuga L., where it was first noticed by Dr. Gray, in 1831. Ledyard, 1827, (*Herb. J. J. Thomas.*) Near Ludlowville Sta. and north to Lake Ridge Pt., where it is abundant. Union Springs. Canoga. Summit Marsh.

"A perennial in muddy or dry places often 2°-3° feet high. Scabrous with short appressed or glandular hairs especially upon the leaves and upper stem. Leaves thin, rather broadly lanceolate or cordate at the base, 4′-7′ long, on short stout petioles ($\frac{1}{2}$′-1′ long) from near the base of the naked sheath. Flowers and fruit nearly as in *P. amphibium*, L. but spikes more elongated (1′-3′) often in pairs."

"New Eng.—Texas and westward. It includes most of the *P. amphibium*, var. *terrestre* of American botanists, and is the *P. amphibium*, L., var. (?) *Muhlenbergii, Meisn.* in DC. Prod. 14. p. 116." (See *Watson, l. c.*)

766. **P. Hartwrightii,** Gray. (*Proc. of Amer. Acad.* VIII, p. 294. 1870.)
(**C.** by Dr. Wright.)
Sphagnum swamps and sedgy marshes; frequent. Aug.-Sept.
Round Marshes. Near pond S. E. of Chicago Sta. Locke Pond. Summit Marsh. Newton's Pond, W. Junius. "Penn Yan," (*Dr. S. Hart Wright,* who called Dr. Gray's attention to the species.)

"Strigose-hirsute or smoothish, stem less than a foot high, erect, striate, equally leafy even to the top; leaves broadly lanceolate, acute or obtusish at each end, short-petioled. Ochræ leaf-like to the middle, hypocrateriform, limb foliacious, reticulate, repand, setose-ciliate. Peduncle erect not glandulose, bearing mostly a solitary, dense, cylindrical spike; bracts surpassing the pedicels; periogonium rose-colored without glands. Stamens 5. Style deeply 2-cleft. Sedgy bogs, N. Y. to Michigan and Wyoming." (*Gray, l. c.*) Dr. Wright thinks its affinities are with *P. amphibium*, L. instead of *P. Careyi, Olney.*

767. P. Virginianum, L.   (H. and C.)
Shaded soil ; not common. Aug.-Sept.
"Lake Marshes" (*Kellermann*, 1873.) probably near the Indian
Spring. Near Negundo Woods. Wild bank beyond the Fleming
S. H. Renwick Farm.

768. P. aviculare, L. KNOT-GRASS.   (H. and C.)
Roadsides ; frequent. Aug.-Oct.
A form on the sands at Sheldrake Point, has narrow glaucous
leaves and large shining stipules, thus resembling *P. maritimum*,
L.

769. P. erectum, L., (*P. aviculare*, L., var. *erectum*, Roth,)
                   (H. and C.)
Roadsides, borders of rich fields ; frequent. Aug.-Oct.

770. P. arifolium, L. TEAR-THUMB.   (C. by Dr. Wright.)
Marshes ; infrequent. Aug.-Sept.
Marshes near Inlet and Fall Cr. Ledyard, 1827, (*Herb. J. J.
Thomas.*) Ringwood. Near Freeville. Summit Marsh. Cayuga
Marshes.

771. P. sagittatum, L. TEAR-THUMB.   (H. and C.)
Low grounds ; abund., especially on the Inlet Marshes. Aug.-
Sept.

772. P. CONVOLVULUS, L. BLACK-BINDWEED.   (H. and C.)
Cultivated ground ; common. Aug.-Oct.

773. P. dumetorum, L. CLIMBING-BUCKWHEAT.
Marshes and river-banks ; frequent. Aug.-Sept.

774. P. dumetorum, L. var. scandens, Gray, is also frequent.
                   (C. by Dr. Wright.)

FAGOPYRUM, Tourn.

F. ESCULENTUM, Mœnch. BUCKWHEAT.   (H. and C.)
Fields and copses ; doubtfully permanent in any place. June-
July.
Fall Cr. Abund. in a piece of woods E. of McLean, 1884 and 1885.

ARISTOLOCHIACEÆ.

316. ASARUM, Tourn.

775. *A. Canadense, L. WILD GINGER.   (H. and C.)
Rich soil, woods, and banks of ravines ; frequent. Apr.-May.

317. ARISTOLOCHIA, Tourn.

776. A. CLEMATITIS, L.
Union Springs, on a bank S. E. of the depot. June.
The plants have probably escaped from some garden, formerly near
the locality. First seen in 1874. They have now doubled the area
occupied then, spreading from the root and forming a dense vigor-
ous growth. Fruit occurs rarely.

PIPERACEÆ. (including SAURURACEÆ.)

318. SAURURUS, Linn.

777. S. cernuus, L. LIZARD'S-TAIL.   (H. and C.)
Wet borders of marshes ; local but not rare. July-Aug.

Indian Spring Marsh. Dryden L. valley. Locke Pond. Abundant on the borders of Cayuga and Montezuma Marshes.

LAURACEÆ.

319. SASSAFRAS, Nees.

778. *S. officinale, Nees. SASSAFRAS. (H. and C.)
Woods, chiefly on the hills ; frequent. May.
In or near all our ravines, but usually small. A large tree grows
beyond the Fleming S. H. and middle-sized ones near upper part of
Geer's Gulf. Frequent on Eagle Hill, in Dart Woods, Turkey Hill
Woods, and woods on the hills of Caroline and Danby.

320. LINDERA, Thunberg.

779. °L. Benzoin, Meisner. SPICE-BUSH. (H. and C.)
Low grounds ; common. Apr. 20–May 10.

THYMELACEÆ.

321. DIRCA, Linn.

780. *D. palustris, L. LEATHERWOOD. MOOSE-WOOD.
(H. and C.)
Rich woods ; infrequent. Apr. 20–May 10.
Case. Creek. Six Mile Creek on Beechwood Flats. Dryden-Lan-
sing Swamp. Round-Marsh Woods. Swamps of Groton. North
of Cayuta L.

322. DAPHNE, Linn.

781. D. MEZEREUM, L. The MEZEREUM was detected growing spontaneously by the road from Trumansburg to Taughannock, 1884 (J.
L. Coville.) Also a single shrub on the east shore of Cayuta L.
1885. (F. V. Coville.) Hillside near Groton, 1886, (W. Morton.)

ELÆAGNACEÆ.

323. SHEPHERDIA, Nutt.

782. S. Canadensis, Nutt. SHEPHERDIA. (H. and C.)
Cliffs and rocky slopes, where it is frequent. May 1–15.
Rare in Case. Cr. in Six Mile below Well Falls ; in Fall Creek
below Mirror and Ithaca Falls. Common on E. shore of Cayuga
Lake ; less common on west shore ; Taughannock and elsewhere.

SANTALACEÆ.

324. COMANDRA, Nutt.

783. *C. umbellata, Nutt. COMANDRA. (H. and C.)
Dry woods with low *Vacciniums* and *Ceanothus ;* common. May
10–June.

EUPHORBIACEÆ.

325. EUPHORBIA, Linn.

784. E. maculata, L. CREEPING SPURGE. (C.)
Fields and roadsides ; common. July–Oct.

785. E. hypericifolia, L. (C.)
Fields in dry soil ; frequent. July–Sept.

[E. PLATYPHYLLA, L. at Penn Yan. *S. Hart Wright*]

786. E. ESULA, L.
Roadsides; scarce. June-July.
Groton, 1875; again, 1884.

787. E. CYPARISSIAS, L. CYPRESS SPURGE.
Gravelly roadsides; occasional. May-June.
Escaping from the Valley Cemetery. Near R. R. beyond Enfield Creek. On the hill beyond Enfield ravine, abundant and not near any dwelling. By R. R. east of Varna.

788. E. PEPLUS, L.
Scarce; streets of Ithaca; Buffalo St.; E. State Street; Aurora Street and Mill Street. Grounds of the Pres. White place.

326. ACALYPHA, Linn.

789. **A. Virginica**, L. (H. and C.)
Fields, etc; common. Aug.-Sept.

[MERCURIALIS ANNUA, L., near Penn Yan. *Dr. Wright.*]

CERATOPHYLLACEÆ.

327. CERATOPHYLLUM, Linn.

790. **C. demersum**, L. HORNWORT.
Aquatic; frequent in all ponds and lakes, but not fruiting.

CALLITRICHACEÆ.

328. CALLITRICHE, Linn.

791. **C. verna**, L. WATER-STARWORT. (H. and C.)
Ditches and shallow water; frequent. Blooms all the year.

792. **C. heterophylla**, Pursh. WATER-STARWORT.
Muddy places and shallow water; frequent. All the year.
Forma brevifolia, and *forma linearis* are both common.

URTICACEÆ.

329. ULMUS, Linn.

793. *U. fulva, Michx. SLIPPERY ELM. (H. and C.)
Rocky banks and ravines; frequent. April 20-May 10.
The Campus Brook. Fall Cr., north side. Six Mile Creek, below the mill and elsewhere.

794. *U. Americana, L. WHITE ELM. (H. and C.)
Rich soil or low grounds; common. Apr. 15-30.
The beautiful form, with slender pendulous branches is frequent, especially along Cayuga Lake; a fine example near Taughannock landing; also near Kidder's Ferry and north; at Indian Falls, Ludlowville. The "Big Elm" at the Elm Tree tavern in McLean is of this form. Old first-growth trees of this form, growing along the borders of swampy woods in Groton, Genoa, and Dryden, exhibit peculiarities of their own; the trunk being usually very tall and straight, with a crown of short, stout, angular branches, and long, pendulous, ultimate branchlets at the summit, clusters of the slender branches of recent growth hanging from sides of the main trunk at intervals. They form the most striking feature in a distant view of these woods.

795. **U. racemosa**, D. Thomas. CORKY ELM. (**C.**)

Declivities and shores ; also low, rich woods ; frequent. May 1–10. Six Mile Creek, and many of our ravines, especially on sunny slopes. Cayuga L. shore. South of Danby village. Dryden. Conspicuous east of McLean, as tall trees with short branches, thus presenting a columnar aspect. At McLean R. R. Sta. is a large tree of this species ; and a remarkable specimen is on the alluvial soil west of Larch Meadow. The latter is fifteen to twenty meters tall ; its trunk in descending, forking about a meter and a quarter above the ground, the two divisions entering the ground over a meter apart. These divisions are much flattened, and from one-half to three-quarters of a meter in the greatest diameter.

330. CELTIS, Tourn.

796. **C. occidentalis**, L. HACKBERRY. (**C.**)

Declivities and bottom lands ; infrequent. Apr. 25–May 15. Fall Breek, below Ithaca Fall. Renwick-Farm slope, and near Pleasant Grove Brook. Negundo Woods and lane west. Neguæna Creek. W. Hill.

331. MORUS, Tourn.

797. * **M. rubra**, L. RED MULBERRY. (**H.** and **C.**)

Ravines and hillslopes ; not rare. June. By road Fall Cr. to the "Nook." Renwick Farm. Below Green-Tree Falls. Cayuga Lake ravines ; as Salmon Cr., Franklin's, Big Gully, etc. In Danby and Enfield. Not seen in the low woods of Dryden.

798. M. ALBA, L. WHITE MULBERRY. (**H.** and **C.**)

Ravines and hillslopes ; scarce. June. Case. Creek, below Glen Pond. Six Mile Creek, near D. L. & W. railroad, and near the Ferris Brook. Near road beyond the "Nook." North of Taughannock ravine and elsewhere.

332. URTICA, Tourn. NETTLE.

799. **U. gracilis**, Ait. (**H.** and **C.**)

Low grounds and waste places ; common. July–Aug.

800. * U. DIOICA, L. (**H.** and **C.**)

Rich shaded soil ; not common. Aug. Negundo Woods. Near road to Buttermilk Falls.

333. LAPORTEA, Gaudichaud.

801. **L. Canadensis**, Gaud. WOOD NETTLE. (**H.** and **C.**)

Rich woods ; frequent. July–Aug.

334. PILEA, Lindl.

802. **P. pumila**, Gray. RICH-WEED. (**H.** and **C.**)

Moist shaded soil ; common. July–Sept.

335. BOEHMERIA, Jacq.

803. **B. cylindrica**, Willd. FALSE NETTLE. (**H.** and **C.**)

Shaded soil. Frequent, especially on the marshes. July–Sept.

336. PARIETARIA, Tourn.

804. **P. Pennsylvanica**, Muhl. PELLITORY.

Dry places beneath cliffs, where it is frequent. June–Aug.

Occurs in all our ravines, and near the cliffs of the lake shore. " Ithaca," (*Herb. Sartwell.*)

337. CANNABIS, Tourn.

805. C. SATIVA, L. HEMP. (H. and C.)
Rich, waste soil ; scarce. July–Aug.
Near Cascadilla Place. Six Mile Creek, near Ithaca. Forest Home.

338. HUMULUS, Linn.

806. H. Lupulus, L. HOP. (H. and C.)
Along banks of streams, and in swamps ; scarce. July.
Island in Beebe Pond. Near mouth of Coy Glen. Near Fall Creek, east of Freeville. Round Marsh and several places in the interior of Beaver Creek Swamp, where it is manifestly native.

PLATANACEÆ.

339. PLATANUS, Linn.

807. * P. occidentalis, L. BUTTONWOOD. SYCAMORE. (H. and C.)
Woods, especially in alluvial soil ; frequent. June.
Along Cascadilla and other creeks. These trees—some of them of great size—formerly gave to Negundo Woods the aspect of the western bottom-land woods.

JUGLANDACEÆ.

340. JUGLANS, Linn.

808. * J. cinerea, L. BUTTERNUT. (H. and C.)
Ravines and rich uplands ; frequent. May 15–30.

809. *J. NIGRA, L. BLACK WALNUT. (H. and C.)
Roadsides, fields ; rarely in ravines; scarce. May 15–30.
Scattered trees have been observed in about a dozen localities within limits, but in none is the species clearly indigenous except possibly a station along a creek in Fayette. The tree nearest the Campus is on Factory street.

341. CARYA, Nutt.

810. *C. alba, Nutt. SHAGBARK-HICKORY. (H. and C.)
Woods and slopes ; common. May 15–June 10.
Campus, northwest of the chapel, and west of the Engineering Building ; and elsewhere, in woods.

811. C. microcarpa, Nutt. WHITE HICKORY. (Dr. Wright in C.)
Hillsides ; not uncommon. May 15–June 10.
Frequent on the Campus, (excellent examples in front of Dr. Law's residence.) Cayuga L. west shore, in Stevens' woods; also near Crobar Point and in Wayne's Woods near Canoga. Woods near Eagle Hill and Snyder Hill.
This species is reduced to *C. alba, Nutt,* by some. In this region it is clearly distinct from that species; its affinities being with *C. tomentosa,* Nutt., and *C. porcina,* Nutt. It also agrees clearly with a large series of specimens of *C. microcarpa,* at the U. S. G. Herb. at Washington, received from different parts of the Union, and is not essentially different from Nuttall's figure and description. I re-retain it as a distinct species, because its suppression would be confusing to any student of our local plants.

812. **C. sulcata**, Nutt.   WESTERN SHAGBARK-HICKORY.

Low bottom-lands, or borders of marshes ; scarce.   June 1-15.

Borders of Cayuga Marshes, west of Cayuga Bridge, (a few trees.)
Occasional in low woods in Junius.   (Mr. Hankenson sends me specimens apparently of the same form, from borders of streams near
Lyons and Newark, Wayne Co.)   All the foregoing are not strik-
ingly different from *C. alba*, *Nutt.*, at first glance, but are undenia-
bly *C. sulcata*.   More striking specimens—differing in some re-
spects from the type—are those in the alluvial swamp, east side, and
not far from the mouth of Owasco L. inlet.   These were the earliest noticed—(July, 1881.)   *C. sulcata*, Nutt., has not been credited
to N. Y. hitherto, although Dr. Torrey in the N. Y. Flora, II., says :
" it is probable that *Carya sulcata* is a native of N. Y., although I
have not found it within the limits of the state."   The habitat ;
the very shaggy bark ; the 7 hairy leaflets on the ordinary branches
and 9 leaflets on young shoots ; the winter buds—approaching *C.*
alba ; the bark of twigs –light-colored and less firm and tough,
than *C. alba ;* the exocarp, oblong and of light, porous character
always different from *C. alba ;* the thick-walled rather large, coffee-
colored nut, are the most important characters distinguishing this
species.   The attention of local collectors in Central N. Y. being
called to this form, probably it will be found not uncommon.   Prof.
Sargent writes that there is a large and valuable nut sometimes sold
in the markets in the "Genessee country" under the name of
" *King Nut*," which may belong to this species.   It would be interesting to know what it is, and how widely the name prevails.

813. **C. porcina**, Nutt.   PIGNUT HICKORY.   (C.)

Upland woods and dry banks ; common.   May 15-June 10.

Especially abundant near the shore of Cayuga L., in woods.

814. ***C. amara**, Nutt.   BITTERNUT HICKORY.   (H. and C.)

Ravines, and rich soil ; frequent.   June.

Near Renwick.   Banks of Six Mile Creek ravine.   Woods and
banks near Freeville and McLean.   Cayuga L. ravines ; at Crobar
Pt., Paine's Creek, Aurora, and near Levanna and elsewhere.

MYRICACEÆ.

342. MYRICA, Linn.

815. **M. Comptonia**, C. DC.   (*Comptonia asplenifolia*, Ait.)   SWEET-
FERN.   (H. and C.)

Gravelly soil on the higher hills ; not common.   Apr. 20-May.

Hills on both sides of White Church valley.   Danby, on Ball Hill
and ridges east of W. Danby.   High hills south of Saxon Hill.

816. **M. Gale**, L.   SWEET GALE.

Cold sphagnum swamps ; rare.   Apr., May.

Locke Pond, above and below the bridge, (Savannah, N. Y., *Sart-*
well, H. and C.)

817. **M. cerifera**, L.   BAYBERRY.   Junius, in the sphagnum about
Newton's and Lowery's Ponds.   "Junius," (*Herb. Sartwell.*)

BETULACEÆ.

343. BETULA, Tourn.

818. **B. lenta**, L. BLACK BIRCH. SWEET BIRCH. (H. and C.)
Woods and ravines; frequent. May 10–25.

819. **B. lutea**, Michx. YELLOW BIRCH. (H. and C.)
Ravines and swampy woods; frequent. May 10–25.
[*B. alba, var. populifolia, Sp.* and *B. papyracea, Ait* are in Sart-
Well's Cat. of 1844; "probably from Wayne Co.," *Wright.*]

344. ALNUS, Tourn.

820. **A. incana**, Willd. SPECKLED ALDER.
Swamps and near streams.; the common alder. Mar. 20–April.

821. **A. serrulata**, Willd. SMOOTH ALDER. (H. and C.)
Shores; scarce, always in vicinity of Cayuga L. Mar. 20–April.
Indian Spring Marsh, and marsh along north end of Cayuga St.
Banks of Inlet, and near Light-house. South of Ludlowville Sta.
Crobar Pt. Lockwood's Flats. More or less abundant west side of
Cayuga Marshes.

CUPULIFERÆ.

345. CORYLUS, Tourn.

822. **C. Americana**, Walt. AMERICAN HAZEL-NUT. (H. and C.)
Hillsides and by streams; frequent. Mar. 20–April.
Chiefly in the Negræna valley; along the creek near the McGraw
barns. On Larch Meadow; near Coy Glen and the Valley Cemetery.
Casc. ravine, north side, also south of Eddy Dam, above the path.

823. **C. rostrata**, Ait. BEAKED HAZEL-NUT. (H. and C.)
Hillslopes and ravines; rather common. Mar. 20–Apr.
Found in or near all our ravines, in copses near the lake, and on
the hills in Danby and Caroline as well as in the lower valleys.

346. OSTRYA, Micheli.

824. **O. Virginica**, Willd. IRON-WOOD. DEER-WOOD. (H. and C.)
Woods and ravines; common. Apr. 25–May 15.

347. CARPINUS, Linn.

825. **C. Caroliniana**, Walt. (*C. Americana*, Michx.) BLUE BEECH.
(H. and C.)
Swampy woods and ravines; common. May 1–20.

348. QUERCUS, Linn.

826. *Q. alba, L. WHITE OAK. (H. and C.)
Woods; common. May 25–June 10.

827. Q. macrocarpa, Michx. BUR-OAK. (H. and C.)
Low grounds and sunny banks of ravines; infrequent. May 25–
June 5.
Near Renwick. Below Buttermilk Falls. Summit Marsh. North
of Freeville. Cayuga Marshes and elsewhere.
A form with shallow, flat acorn-cups occurs on Lake Ridge Point.

828. Q. bicolor, Willd. SWAMP WHITE-OAK. (H. and C.)
Low grounds; frequent. May 25–June 5.

Field south of Indian Spring Marsh, and scattered over the Ithaca plain. Occasional in the low woods throughout our territory and extending to Cayuga and Montezuma Marshes.

829. * Q. Prinus, L. ROCK-OAK. CHESTNUT-OAK.   (H. and C.)
Chiefly on banks of ravines and dry hills; frequent. May 15-30. Most of our specimens have the large nuts, with turbinate cups, and leaves usually broad. Forms with narrow leaves and small nuts occur at White Church, Fall Cr., Snyder Hill and elsewhere. A form at White Church has nuts small borne on peduncles 2-3 cm. long.

830. Q. prinoides, Willd. (Q. pumila, Michx.; Q. Prinus, L., var. humilis, Marsh.) 1. This, the dwarf form, occurs very sparingly along the east shore of the lake, north of the High Cliffs between King's Ferry and Willet's. (Penn Yan, Herb. Sartwell.)

831. 2. The arboreal form, (Q. Prinus, L., var. acuminata, Michx.; Q. Muhlenbergii, Engelm; Q. Castanea Muhl.,) is also rare, on limestone or in its vicinity. May 20-June 1.

On first ledge of Tully limestone, south of Shurger's Glen, a few trees. Big Gully, in ravine above the crossing of the second road from the lake, a considerable number of small or middle sized trees. Our trees have the characteristic bark, leaf and nut of this species. I follow Prof'r Sargent in joining these two seemingly distinct forms, as he has had a wide range of specimens and observations on which to base his judgment.

832. Q. coccinea, Wang. SCARLET OAK.   (H. and C.)
Dry hills; infrequent. May 15-25.
Frequent on the campus, (good examples at the north entrance.) Sparingly at Six Mile Creek Narrows, and near S. S. 420, on South Hill. North of Salmon Creek, and of Lake Ridge, otherwise scarce near the lake shore. Woods N. W. of Pleasant Grove Cemetery. Near Junius Ponds.

833. Q. tinctoria, Bartram. (Q. coccinea, Wang., var. tinctoria, Gr.) YELLOW OAK.   (H. and C.)
Woods, especially on hills; common. May 15-20.

834. Q. rubra, L. RED OAK.   (H. and C.)
The common oak of the drier woods and lake shore. May 15-25.
A good example occurs by the Campus Brook, at the Sage Coll. bridge. The leaves of this oak often change to a fine bronze and brownish scarlet, as well as yellow, in the Autumn. Apparent hybrids with Q. coccinea occur on the Fiske-McGraw grounds.

349. CASTANEA, Tourn.

835. C. vulgaris, Lam., var. Americana, A. DC. (C. vesca, L., var. Americana, Michx., Man., p. 455.) CHESTNUT.   (H. and C.)
Woods and banks of all the ravines. July 10-20.
Especially abundant in Dart Woods, woods from Turkey Hill to Ringwood; and on the hills about White Church.

350. FAGUS, Tourn.

836. * F. ferruginea, Ait. BEECH.   (H. and C.)
Woods and ravines; common. May 25-June 10.

According to Mr. Lewis of W. Hill, "Red Beech" is the form growing in drier soil and somewhat dwarfed; "White Beech," the form—larger, and of more rapid growth—occurring in richer soil.

SALICACEÆ.

351. SALIX, Tourn. WILLOW. OSIER.

837. **S. nigra,** Marshall. BLACK WILLOW. (H. and C.)

Along streams and low shores ; common. Flowers May 20-June 1. Casc. Cr. Especially abundant Six Mile Cr. and Neguæna Cr. and at the head of Cayuga Lake, where it grows in great masses. Its slender bright, green leaves, borne in great abundance help to relieve the flat, shore lines more than any other willow. A form with thick. elliptical leaves more woolly on the petiole and branchlets is found near Canoga and elsewhere. The form known as var. *falcata* is occasional.

838. **S. amygdaloides,** Andersson. (*Salices Boreali-Americanæ,* in *Proc. Amer. Acad. of Arts and Sci.,* Vol. IV.) (*S. nigra,* Marsh. var. *amygdaloides,* Anderss. *Mon. Sal.* p. 21.) PEACH WILLOW.

"Leaves broadly lanceolate, 3'-6' long, $\frac{3}{4}$-1$\frac{1}{2}$' wide, with a long. tapering point, glaucous beneath, closely serrate, petioles long and slender, stipules minute and very early deciduous. Aments, leafy peduncled, elongated-cylindrical, pendulous ; the fertile, when in fruit, lax, 3'-4' long, $\frac{1}{4}$' thick ; scales in the male ament ovate. villous with crisp hairs, in the female narrower, somewhat smooth, fugacious : Capsules globose-conical, glabrous, long-pedicelled ; Style very short or obsolete ; stigmas notched," (*Bebb, in the Botany of the Wheeler Survey, West of the 100th Meridian,* p. 240.) Andersson says the flowers are triandrous, but ours usually have 5 or 6 stamens. The older trees attain a height of 9-12 meters and the bark resembles that of *S. nigra,* but the large lanceolate leaves. whitish below, and the reddish young shoots give it an entirely different aspect. From form of leaf and color of twig it is well-named the "Peach-leaved Willow." Its twigs are somewhat tough at the base while those of *S. nigra* disarticulate very readily. Professor Sargent in "*Forest Trees of Nor. Amer.,*" (*Census Rep.,* Vol. IX. 1880,) gives the eastern limits of distribution, as Wayne Co., N. Y. We have observed it in stations 30 or 40 miles east of this range.

Low grounds, streams and shores; abundant. Flowers May 10-30. Casc. Cr. north of Eddy Pond. Six Mile Cr. Neguæna valley,— abund. Cayuga Lake, whole length. Cayuga Marshes. Montezuma Marshes. Near Waterloo. (Watkins, Seneca L.) Venice, west of Cascade. Scarce near Freeville and Cortland.

839. **S. lucida,** Muhl. SHINING WILLOW. (H. and C.)

Streams and shores ; frequent. May 20-30. Casc. Cr. below Eddy Pond. Frequent along the low points of Cayuga Lake. In Cayuga Marshes. Near Freeville, McLean, Beaver Cr. Dryden Lake valley, Bear Swamp, and elsewhere.

840. **S. lucida,** Muhl. var. —— with beautiful shining, coriaceous. very finely serrate leaves, larger, light-brown pods, and flowers and fruits very much later than in the type, occurs in the Round-Marshes. I have also collected it in Bergen Swamp, N. Y., from

which place Mr. Bebb has received it. as well as sparingly from N.
J., Ohio, and Mich. Flowers from June 10-30 and matures fruits
slowly, the writer obtaining pods still in excellent condition, Sept.
9. 1880.

841. S. FRAGILIS, L. BRITTLE WILLOW.   (H.)
   Form, (cultivated) ; "near true *S. fragilis*," according to Mr.
Bebb. May 10-20.
   Planted near the south barn on Univ. Farm. Near Hibbards Corners. Near Fleming S. H. King's Ferry Sta. Atwater's Sta., etc.
This form is not pure *S. fragilis*, L. but probably the result of a
crossing several times over with already mixed forms of *S. alba*, L.
in which result *S. fragilis*, L. decidedly predominates. It is extensively propagated from cuttings and planted as a wind-break—passing under the name of "Black Willow." Specimens which cannot
be clearly distinguished from the cultivated are found on Fall Cr.
Marsh, on Myers Pt. and elsewhere growing spontaneously.

842. S. fragilis, L. × S. alba, L.
   Spontaneous forms with yellowish twigs from Negundo Woods
and by the R. R. near the woods ; also by the R. R. east of Freeville and east of McLean ; Cayuga Lake shore at Taughannock, on
Myer's Point, Lockwoods Flats and elsewhere, are here included,
and are pretty clearly hybrids between the above so-called species,
having characteristic traces of their flowers and leaves.

843. *S. ALBA, L. var. VITELLINA, Koch.— (? form. Probably a hybrid derived from the true var. *vitellina*, Koch. and *S. fragilis*, L. Its
branchlets are usually yellow.) YELLOW WILLOW.   (H. and C.)
   Planted to protect creek and R. R. embankments. May 10-20.
   Willow Pond. Willow Ave. Ithaca along Six Mile Creek, and
elsewhere. Also frequently wild along streams and shores. Always
pistillate. The hybrids mentioned above, between *S. fragilis* and
S. alba are no doubt derived on the *S. alba* side from the present
form, which has been long in cultivation ; but the history and origin of the American hybrids, between these two species, are admit-
ted by the best authorities to be involved in the greatest obscurity.

844. S. alba, L. var. vitellina, Koch, × S. lucida, Muhl, (probably.)
   Negundo Creek ; marshes north of Ithaca, along the Cayuga So.
R. R.; and Cayuga L. to Union Springs ; not uncommon. May 10-20.
   An arborescent form of luxuriant growth, sometimes its trunk and
branches are quite yellow, sometimes gray-green. Its wood is
everywhere brittle. Both staminate and pistillate plants occur ; and
when in flower, the former are as striking as the best marked species of willow. Catkins are very abundant, bright yellow, 6-7 cm.
long, and fill the air with their peculiar balsamic fragrance. The
stamens,—usually 4-5 in each flower,—leaves and stipules suggest
S. lucida, while the long catkins, color of twigs in many, and the
habit of the plant, are like *S. alba-vitellina*. (Also grows in
Wayne Co., N. Y., *E. L. Hankenson*.)

S. ALBA, L. var ARGENTEA. —— distinguished by its silky-canescent
leaves and young twigs ; and S. ALBA, L. var. VITELLINA, Koch, are

in cultivation on the Fiske-McGraw grounds. Cuttings of these and S. ALBA, L. (*pura,*) have been transferred to Cascadilla Cr. and Woods.

Several other forms, each with peculiar characteristics, are found in different places, chiefly near the lake. These appear to be hy⁻ brids derived from the *fragiles* group, (with the influence of the *nigra* group apparent in some,) but their exact position cannot be determined at present.

845. *S. BABYLONICA, Tourn.   WEEPING WILLOW.   (H).
Planted, and rarely escaping along the lake, (probably from wind-fall twigs) as at Myers Point.   May 10–20.

846. S. longifolia, Muhl.   LONG-LEAVED WILLOW.
Sandy and rocky shores ; not common.   May 15–June 10.
Fall Cr. below Island Fall, (*Miss I. S. Arnold.*)!   North of Esty Glen.   Myers Pt. and near Ludlowville Sta.   Lake Ridge Point.   In several places between King's Ferry and Lake Ridge.   Mouth of Paine's Creek.   Lockwood's Flats.   Farley's Point.   Canoga.   Often blooming in Aug. on the lake-shore, where the twigs are badly dis-torted by a fusiform insect-gall.

847. S. discolor, Muhl.   GLACOUS WILLOW.   (H. and C.)
Wet places, or even rocky banks ; common.   Apr. 15–30.
The earliest willow to bloom, and one of the most variable.   An-thers sometimes salmon-pink, sometimes wholly or partially trans-formed into ovaries.   Young catkins sometimes 10–12 cm. long and cylindrical, while in specimens on Summit Marsh, they are 1½ cm. and globular.   The latter specimens are peculiar in their yel-low-brown bud scales, coriaceous, shining, almost cuspidate leaves, and dwarf stems.

848. S. rostrata, Richardson. (*S. livida*, Wahl. *var. occidentalis*, Gr., Man., p. 464.)
Ravines, lake-shore and marshes ; frequent.   Apr. 20–May 10 ; or later in sphagnum marshes, and rarely a second time, in Aug.
It occasionally becomes a low tree 15–25 cm. in diameter.   Bulb-ous insect-galls in the twigs are frequent.

S. CAPREA, L. (*forma pendula*) characterized by its umbrella-like mass of drooping branches, and its broad, woolly leaves, is frequently planted in town, and is allied to the preceding species.

849. S. myrtilloides, L.   MYRTLE WILLOW.
Sphagnum swamps ; rare.   May 20–30.
Freeville, north of the village, only the pistillate plants.   "Lodi Seneca Co., *Dr. Folwell*" (See *Flora of N. Y.* II p. 213.) Junius Seneca Co. (*Sartwell H. and C.*)

850. S. petiolaris, Smith.   (H. and C.)
Low grounds and marshes ; frequent.   Apr. 25–May 15.
Fall Cr. on the Circus Common, and in the Marsh.   Lake Ridge Point.   Cayuga Marshes, abundant.   Summit Marsh, a dwarf form with short catkins.   This species, especially the staminate plant, be-fore the leaves appear, is difficult to distinguish from *S. sericea*, Marsh.   The writer has found without exception in testing a large

number of cases, that the twigs do not disarticulate at the base, i. e., they are "tough," while the twigs of *S. sericea* have long been known to be "brittle," (See *Marshall, Arbustum Amer.* 1785. p. 140; and *Pursh, Flora Amer. Sept.* 1814, II, p. 616.)

851. **S. sericea**, Marshall. SILKY WILLOW. (H. and C.)
Low grounds; common. Apr. 20-May 10.
Case. Cr. below Eddy Pond. Fall Cr. and the borders of the lake-marshes. Abund. at Freeville, Round Marshes, Beaver Cr. and else-where. Dwarf forms with beautifully silky leaves at Summit Marsh. The fusiform galls on the twigs very abundant and broader than in *S. longifolia*.

852. **S. humilis**, Marshall. LOW WILLOW. (H. and C.)
Woods along the glens, and frequent on the tops of the highest hills. Apr. 15-30.
Case. Woods. Fall Cr. Coy Glen. South Hill. High hills of Dryden, Caroline, Danby and Newfield. The delicate pink of the young anthers is very characteristic.

853. **S. cordata**, Muhl. HEART-LEAVED WILLOW. (H. and C.)
Low grounds and shores; common. Apr. 20-May 10.
Variable in the appearance of pistillate catkin and scales, and in leaf; also hybridizing freely. This species, *S. humilis* and *S. dis-color* often produce galls in the form of cones at the ends of the branches.

854. **S. cordata**, Muhl. × **S. sericea**, Marsh. (*S. myricoides, Muhl:—Salix cordata*, var. *myricoides*, of Gray's Man., p. 464, probably.)
Fall Cr. on the Circus Common. Near the head of the lake. By the R. R. south of Ithaca, and elsewhere; frequent.

855. **S. cordata**, Muhl. × **S. petiolaris**, Marsh.
West Danby valley. Certain forms from the Fall Creek Common also seem to belong here.

856. **S. candida**, Willd. HOARY WILLOW. (H. and C.)
Sphagnum marshes or cold swamps; rare. Apr. 25 May 10.
Fleming Meadow. Locke Pond. Junius, (*Sartwell H. and C.*) near Newton's and Lowery's Ponds. Larch Swamp, S. W. of Sa-vannah, N. Y.

857. **S. incana**, Schrank. × **S. cordata**, Muhl. ? nov. hyb.
One cluster, spontaneous, on East hill, opposite residence of E. G. Jayne. April 25-May 10.
Detected, 1884; flowers collected May 2, 1885.
Stem, shrubby, dark, reddish-brown, branchlets somewhat wool-ly-canescent, not disarticulating at the base. Buds oblong, brown or reddish brown, outer scales of the flowering buds separating into parallel plates as in *S. cordata*, Muhl, and *S. candida*, Willd. *Leaves* linear-lanceolate, numerous, 3-8 cm. long, 4 6 mm. broad in the middle, tapering to a rather acuminate apex, and below into short petiole, margin sharply and regularly glandular-serrulate, at first white tomentose all over, soon nearly smooth and rather dark green above, glaucous and more or less tomentose below. The

foliage altogether of a gray or silver-green aspect, and does not blacken in drying. *Stipules* narrowly lanceolate or almost setaceous, somewhat glandular-denticulate when young. *Catkins*— pistillate, 2-3 cm. long, (or 3-4 cm. when mature,) slender compact, erect, resembling *S. incana*, Schrank ; abundantly leafy bracted at the base, bracts inclining to be abruptly acute, mucronate or obtuse, even in the mature state. *Ovaries* stalked, glabrous, lanceolate, style slender, stigmas deeply bifid, each branch with two, rarely three, oblong lobes, scale oblong, ⅔ the length of the ovary, smooth on the back but ciliate with numerous long white hairs, whitish or purplish tipped when young, inclining to be acute, especially in the lower part of the catkin. Mature seeds very few.

S. incana, Schrank., a willow from the Alps and Pyrenees, is in cultivation on the grounds of B. G. Jayne ; also of J. D. Carpenter. Buffalo St. Both trees are pistillate. It was clear by comparison, that our hybrid was allied to *S. incana*, and in spite of the presence of occasional stipules I referred the former to the latter as an introduced *variety* of it. Referring the plant to Mr. Bebb, as a final authority, he replied that it was a *hybrid*, of somewhat surprising character. He says : "it cannot have been introduced. There is nothing like it in the European Flora." "It is indeed clearly allied to *S. incana* and just as clearly not *S. incana*. * * * The departure from *S. incana* is wholly in the direction of American and not European forms." He thinks it clearly a hybrid between *S. incana*, Schrank, on one hand, and probably *S. cordata*, Muhl., on the other. (or possibly *S. petiolaris*, Sm.;) although, in the absence of experiment, he does not wish to pronounce finally. He found pistils of *S. incana*, (which I collected here, 1881,) to have been *fertilized*, probably by some American willow, as staminate *S. incana* is not cultivated in this region. Comparing the hybrid with the parent it differs from *S. incana ;* in branchlets tough near the base ; leaves and leafy bracts more abundant at flowering time, shorter more tapering and less woolly beneath when mature, sharply glandular serrate and margins scarcely revolute. (*S. incana* has long, linear, slightly denticulate leaves, woolly beneath with margins revolute ;) stipules present, ovary somewhat longer-stalked ; style longer and slenderer ; stigmas 4 or 5 (instead of 2 or 3) and shorter lobed ; scale elliptic and inclined to be acute (instead of truncate) ciliate with longer, more abundant white hairs. (*S. incana* hybridizes freely with Eur. Willows. (See *And. in DC. Prod.*, XVI, (2) p. 302-305.) It has no near relatives indigenous to the American continent.

858. S. PURPUREA L. PURPLE WILLOW.

Introduced and frequent along the lake-shore. Apr. 20–May 10.

Staminate plants occur near the light-house at the Inlet, at King's Ferry, Aurora, Union Springs, on the Pearson estate by Neguæna Cr. (formerly cultivated there,) near Mud Cr., and Beaver Cr. in Dryden, and in Ulysses and Venice. Pistillate plants are in cultivation in Ithaca, village and cemetery.

352. POPULUS, Tourn.

859. * **P. tremuloides**, Michx. AMERICAN OR QUAKING-ASPEN.
                                                            (**H.** and **C.**)
   Banks of ravines, hillsides and hills ; common. April.

860. **P. grandidentata**, Michx.   (**H.** and **C.**)
   Dry woods and hills ; common, April.

861. * **P. monilifera**, Ait. EASTERN COTTONWOOD. NECKLACE POP-
LAR.   (**C.** by Dr. Wright.)
   Shores and river-banks ; not uncommon, April 25 May 10.
   Fall Creek, Beebe Pond and below Ithaca Fall. Casc. Creek, be-
low the bridge. Near Six Mile Creek. Near Renwick, and fre-
quent along Neguaena Cr., and on the shores of Cayuga Lake.

862. **P. balsamifera**, L. BALSAM POPLAR. (**H.**)   Fl. May 1-10.
   Bank of Tanghannock ravine, on the steep, wooded slope at the
top of the north bank, north of the Falls. The trees are near the
great curve of the ravine bank, and are probably indigenous, resembling in habitat and appearance those seen at Niagara and elsewhere. The trees are small, seven of the larger being only 15-20
cm. in diameter, but appearing like old trees. Wild, but apparently introduced, on the high bank above the C. S. R. R., south of
Willet's Sta. Cultivated trees, (staminate and pistillate) in front of
Mr. Mead's, E. of Judd's Falls ; in meadow near White Church
Sta., and elsewhere.

863. P. BALSAMIFERA, L., var. CANDICANS, Gr.   BALM OF GILEAD,
                                                            (C.)
         Planted occasionally. May 15-30.
   On Heustis St., Ithaca. Danby. McLean. Myers Point. At
Round Marshes, about the borders of Pond Meadow, are thirty or
more trees from ¼ to over ¾ of a meter in diam. From their relative position it is altogether probable that they were planted next
to an old ditch made many years since. But this is only evident
after close inspection, as considerable undergrowth at present sur-
rounds them. The trees are all pistillate. Young trees near, have
probably sprung up from windfall twigs.

864. * P. DILATATA, Ait. LOMBARDY POPLAR.   (**H.**)
   E. of State St., opposite Spring St., Ithaca, but apparently spontaneous, (perhaps from floating twigs,) on Cayuga L. at Myer's
Point, Lake Ridge Point, and near Willet's Sta. (Plants all pistillate.)

P. ALBA, L. (THE ABELE,) near the Cayuga St. bridge over Six Mile
Cr., and on South Hill near the R. R., suckers freely from the
spreading roots. (Plants staminate.)

P. TREMULA, L. EUR. ASPEN, is occasionally planted, on the Campus.

MONOCOTYLEDONS.

HYDROCHARIDACEÆ.

353. ANACHARIS, Richard.

865. **A. Canadensis**, Planchon. WATER SNAKE-WEED. (**H.** and **C.**)
Pools and slow-flowing streams ; common. July-Aug.

354. VALLISNERIA, Micheli.

866. * **V. spiralis**, L. EEL-GRASS. (**H.** and **C.**)
Lakes and their inlets and outlets ; common. Aug.

ORCHIDACEÆ.

355. MICROSTYLIS, Nutt.

867. **M. monophyllos**, Lindl. (**C.** by Dr. Wright.)
In deep cold swamps, rarely in ravines ; scarce. June 20-July.
Swamps near Freeville. Mud Cr. and Beaver Cr. Ellis Hollow
Swamp. Spring Brook on Thacher's Pinnacle, (*O. E. Pearce.*)
Cascadilla ravine, (*Mrs. C. G. Ames,* 1875 ; also within a few years.)
Near Smith's Corner's, north of Cayuta L.

356. LIPARIS, Richard. TWAYBLADE.

868. **L. liliifolia**, Richard.
Near " Ovid, Seneca Co., (*Chickering and Brewer in Sart. Herb.*)
a doubtful specimen.

869. **L. Lœselii**, Richard.
Cold mossy banks, and marshes ; infrequent. June 25-July.
Casc. Creek. Marsh between Fall Cr. and the lake. Freeville.
(*F. C. Curtice.*) Springs near Mud Creek. Ringwood. Summit
Marsh. Cayuga and Montezuma Marshes.

357. APLECTRUM, Nutt.

870. * **A. hyemale**, Nutt. PUTTY-ROOT. ADAM-AND-EVE.
Rich woods ; scarce. June 1-20.
Six Mile Cr., 1872 ! 1879 ! and by O. E. Pearce in 1884. In
McGowan Woods, woods W. of Etna, Negœna valley south of
Larch Meadow, (*F. C. Curtice.*) South of the Dryden-Lansing
Swamp, (*F. L. Kilborne.*) Woods near Varna, (*H. L. Locke* and
C. S. Sheldon.) Near Ludlowville, (*Mr. Lord.*) Ledyard, 1827.
when it was frequent, (*Herb. J. J. Thomas.*) Ovid, N. Y., (*Chickering and Brewer, in Herb. Sartwell.*)

358. CORALLORHIZA, Haller. CORAL-ROOT.

871. **C. innata**, R. Br.
Deep swamps ; rare. May 20-June 10.
Dryden-Lansing Swamp. Ellis Hollow Swamp. Borders of Mud
Cr. Swamp.

[**C. odontorhiza**, Nutt. Yates Co., *S. H. Wright.*]

872. **C. multiflora**, Nutt. (**H.** and **C.**)
Woods ; frequent. June 20-Aug.

359. SPIRANTHES, Richard. LADIES' TRESSES.

873. S. latifolia, Torr.
Wet places, near springs and streams ; not uncommon. June.
Larch Meadow. Six Mile Creek. Casc. Cr., toward Ellis Hollow.
In White Church valley. Near W. Danby, Freeville and else-
where. Junius, (*Herb. Sartwell.*)

874. S. Romanzoffiana, Chamisso.
Springy, mossy places ; scarce. July–Aug. 15.
Not uncommon in the vicinity of Mud Creek, Freeville. Brook-
ton Springs. Summit Marsh, and Newton's Pond, Junius.

875. S. cernua, Richard.
Marshes and wet meadows ; frequent. Sept.
Larch Meadow, 1878, (*Professor Trelease.*)! Fleming Meadow.
Near Mud Creek. Near Beaver Cr. and Round Marsh. Dryden L.
valley. White Church valley and elsewhere. Junius, (*Herb.
Sartwell.*)

876. S. gracilis, Bigel. (H. and C.)
Dry woods, frequently with pines ; frequent. July 15-Aug. 15.
Casc. Cr. woods. Fall Cr. woods and elsewhere.

360. GOODYERA, R. Br.

877. G. repens, R. Br. (C. by Dr. Wright.)
In cool moist woods of Coniferæ ; not rare. July 25–Aug. 25.

878. * G. pubescens, R. Br. RATTLESNAKE PLANTAIN. (H. and C.)
In woods, especially of Coniferæ ; abundant. Aug.
In Casc. Woods. Fall Creek and elsewhere.

879. G. Menziesii, Lindl.
Michigan Hollow, borders of the deep swamp ; not in flower but
leaves and buds evidently of this species. Dr. Gray obtained it in
" Western N. Y."—it is believed from Seneca Co.,—in 1831.

361. ARETHUSA, Gronov.

880. A. bulbosa, L. ARETHUSA.
Sphagnum bogs ; rare. May 20-June 10.
Freeville bog, 1874, (*E. H. Palmer.*)! Junius, (*Sartwell H.
and C.*)

362. CALOPOGON, R. Br.

881. C. pulchellus, R. Br. CALOPOGON. (H. and C.)
Sphagnum marshes ; scarce. June 20–July 20.
Larch Meadow. Freeville. Malloryville. Round Marsh.

363. POGONIA, Juss.

882. P. ophioglossoides, Nutt. POGONIA. (H. and C.)
Sphagnum bogs ; infrequent. June 20–July 20.
Larch Meadow. Freeville. Malloryville. Round Marsh, and
bog S. E. of Chicago Sta.

883. P. pendula, Lindl. (H. and C.)
Rich soil ; rare. Aug.
Sheldrake Point, (*Dr. A. Gray,* 1831.) (Geneva, *Dr. J. Smith in
Sartw. Herb.*

884. **P. verticillata**, Nutt.

High hills, (in our region,) in moist woods; rare. May 20-June 15. Caroline, in woods on top of Bald Hill, and west slope of Taft Hill. Danby, in woods on Ball Hill.

364. ORCHIS, Linn.

885. * **O. spectabilis**, L. SHOWY ORCHIS. (H. and C.)

Rich woods and moist banks; not uncommon. May 20-June 10.

Probably common in this region formerly, as it was in Ledyard fifty years since, according to Professor Thomas. At present it is evenly and sparingly distributed throughout our Cayuga L. basin.

365. PERULARIA, Lindl.

886. **P. virescens**, Gray. (*Habenaria virescens*, Spreng.)

Marshy places; not common. June 20-July 20.

Marsh between Fall Creek and Cayuga L. (*F. C. Curtice.*) Dart Woods. Open meadow, with club mosses, between Bald Hill and Taft Hill, Caroline.

366. HABENARIA, Willd., R. Br.

887. **H. tridentata**, Hook.

Cold swamps; rare. July 15-30.

Freeville bog, (*F. B. Hine*)! Mud Cr. swamp. Some of the Mud Cr. specimens are wholly wanting in spurs to the corolla. "Junius," (*Herb. Sartwell.*)

888. **H. viridis**, R. Br., var. **bracteata**, Reich. (H. and C.)

Ravine N. of Buttermilk Glen?

889. **H. hyperborea**, R. Br. (H. and C.)

Cold swamps and marshy places; frequent. June 15-Aug.

Beside the common form, which is usually out of bloom by the middle of July, there is a form in beech and maple woods, usually dwarf and blooming in the latter part of July and in Aug. Specimens of this form from McGowan Woods were destitute of spurs and the lip resembled that of *H. dilatata*, Gr.

890. **H. dilatata**, Gray.

Cold swamps; less frequent than the preceding. June 15-July.

891. **H. Hookeri**, Torr. (C. by Dr. Wright.)

Woods of mingled Coniferæ and hardwood; frequent. June-July. Case. Woods (*E. H. Palmer*, 1874) also 1883! University Grove formerly, (*Dr. Jordan.*) The finest specimens are in the old woods on the higher hills. A form approaching var. *oblongifolia*, occurs on Thacher's Pinnacle.

892. **H. orbiculata**, Torr. (H. and C.)

Deep, old woods, especially on the hills; infrequent. June 20-July 20.

It occurs in Dart Woods, Turkey Hill Woods, and Rhodes Woods; also in the extensive woods on Taft Hill, Caroline, and Ball Hill, Danby; in Signer's Woods and elsewhere.

893. **H. ciliaris**, R. Br. YELLOW FRINGED-ORCHIS. Junius, (*Sartwell, H. and C.*)

894. **H. blephariglottis**, Hook.   WHITE FRINGED ORCHIS.   Junius.
    (*Sartwell, H. and C.*)

895. **H. lacera**, R. Br.   GREEN FRINGED ORCHIS.
    Swamps and meadows ; infrequent.   July.
    Southeast of South Hill Marsh in several places.   Dryden-Lansing
    Swamp.   Freeville bog. (*F. B. Hine*)! Lowery's Pond, Junius.
    "Junius" (*Sartw. H. and C.*)

896. **H. psycodes**, Gray.   (**H. and C.**)
    Meadows and swamps ; frequent.   Aug.
    Larch Meadow.   Buttermilk ravine.   White Church and else-
    where.

897. **H. fimbriata**, R. Br.   PURPLE FRINGED ORCHIS.   (**H. and C.**)
    Swamps where it is quite common.   June–July 15.
    Specimens with pure white flowers occur in Michigan Hollow
    Swamp.

                    367. CYPRIPEDIUM, Linn.

898. **C. parviflorum**, Salisb.   SMALL YELLOW LADY-SLIPPER.
                                                        (**H. and C.**)
    Woods and swamps ; not common.   May 25-June.
    Two-flowered specimens come from Six Mile Creek below the
    landslide, (*O. E. Pearce*); and from Michigan Hollow Swamp.   The
    ordinary form in Cascadilla Woods, (*Miss Hale.*) near Freeville,
    Malloryville, etc.

899. * **C. pubescens**, Willd.   LARGE YELLOW LADY-SLIPPER.
                                                        (**H. and C.**)
    Woods and swamps ; frequent.   May 25-June.
    University Grove, formerly, (*Dr. Jordan.*) Cascadilla Woods.
    Abundant in the swamps of Freeville, Beaver Cr. and elsewhere.
    Two-flowered specimens are occasional.

900. **C. spectabile**, Swartz.   SHOWY LADY-SLIPPER.   (**H. and C.**)
    Cold marshes or open places in deep swamps where it is frequent.
    June 15.-July.
    Scarce in Larch Meadow, Buttermilk ravine, and Enfield ravine.
    Abundant along Mud Creek, in Beaver Creek Swamps and Michi-
    gan Hollow Swamp.   In Wyckoff Swamp, and Malloryville Marsh.
    The most splendid of our native orchids, this species is seen in per-
    fection in the half shaded sphagnum openings in the middle of
    thick swamps.   The plants are occasionally a meter or more high.

901. * **C. acaule**, Ait.   PINK LADY-SLIPPER.   (**H. and C.**)
    Rich woods, often from moldering trunks of fallen pines and
    hemlocks ; frequent.   June.
    Rare in Casc. Woods, (*O. L. Taylor*)! and Fall Cr. Woods, but
    found in or near all our larger ravines, and frequent in Dart Woods,
    Rhodes Woods, Round-Marsh woods, and those of a similar char-
    acter, throughout our territory.

                    IRIDACEÆ.

                    568. IRIS, Linn.

902. * **I. versicolor**, L.   BLUE-FLAG.   (**H. and C.**)
    Marshes and swamps ; common.   June.

903. I. PSEUDACORUS, L. YELLOW IRIS.
Established in marshy places ; scarce. June 1–20.
Marsh south of the Glass Works. (*Mr. Lord* ; *F. B. Hine*) !
Between Fall Cr. and the lake, (*F. C. Curtice*) ! South of Union
Springs, near the R. R.

369, SISYRINCHIUM, Linn.

904. * S. anceps, Cav. (*S. Bermudiana*, L., of Man., in part.) BLUE-
EYED GRASS.   (H. and C.)
The typical form near Negundo Woods, Freeville, Poplar Ridge,
and elsewhere. The form with leaves of spathe nearly equal and
narrower, is more frequent ; Casc. Cr. (*Dr. Jordan*) ! Fall Cr.,
Farley's Pt., Freeville, Danby, etc. May–June.

905. S. macronatum, Michx. (*S. Bermudiana*, L., var. *mucronatum*,
Man., p. 517.)
Scarce, in Ithaca Cemetery, (*Kilborne.*) !
Possibly both the above species should be united under *S. angus-
tifolium*, Mill. (See *W. B. Hemsley*, in *Jour. of Bot.*, XXII, p.
108.

AMARYLLIDACEÆ.

370. HYPOXYS, Linn.

906. * H. erecta, L. STAR-GRASS.   (H. and C.)
Meadows and grassy slopes ; frequent. June–July.
Casc. Cr. near Eddy Pond. Fall Cr. South Hill and elsewhere.

SMILACEÆ.

371. SMILAX, Tourn.

[* S. rotundifolia, L., is in Dr. Thompson's Cat., (probably a mistaken
identification of the next.) It is in Sartwell's Cat. of 1844, but the
specimen in his herbarium is from Brooklyn, N. Y.]

907. S. hispida, Muhl. GREEN-BRIER.
Woods and swamps ; frequent. June.
Casc. Woods, near the Goldwin Smith Walk. Fall Creek, and
elsewhere.

908. S. herbacea, L. CARRION-FLOWER.   (H. and C.)
Woods ; frequent. June.
Casc. Woods, Fall Cr., north of Beebe Pond and elsewhere.

LILIACEÆ.

372. ALLIUM, Linn.

909. A. tricoccum, Ait. WILD LEEK.   (C.)
Ravines and rich woods ; infrequent. Leaves in May, flowers in
July.
Six Mile Cr. Taughannock ravine. Woods near Mud Cr. and
elsewhere.

910. A. Canadense, Kalm. WILD GARLIC. (C. by Dr. Wright.)
Ravines and fields ; infrequent. May 20–June 20.
Fall Cr. on the island in Beebe Pond, (*Curtice and Kilborne*) ;
below Ithaca Fall. Field near the Salt-well at Aurora and elsewhere.

ORNITHOGALUM UMBELLATUM, L.—STAR OF BETHLEHEM—was found on the island in Beebe Pond, 1880, by Mr. Curtice ; and rarely escapes to roadsides.

373. POLYGONATUM, Tourn.

911. * **P. biflorum**, Ell.　SOLOMON'S SEAL.　　　　　(**H.** and **C.**)
Ravines and woods ; common.　May.

912. * **P. giganteum**, Dietrich.　LARGER SOLOMON'S SEAL.
(**H.** and **C.**)
Scarce ; in Negundo Woods and elsewhere in the valley.

374. SMILACINA, Desf.

913. * **S. racemosa**, Desf.　FALSE SOLOMON'S SEAL.　(**H.** and **C.**)
Ravines and woods ; common.　May 20–June 20.

914. **S. stellata**, Desf.　　　　　　　　　　　　　(**H.** and **C.**)
Rich woods and spring-marshes ; frequent.　May.

915. **S. trifolia**, Desf.　THREE-LEAVED S.
Very rare ; only in the Spruce Swamp on Enfield hill.　May 20
June 20.

375. MAIANTHEMUM, Weber.　TWO-LEAVED SOLOMON'S SEAL.

916. * **M. Canadense**, Desf. (*Smilacina bifolia*, var. *Canadensis*, Gr., Man., p. 530.)　　　　　　　　　　　　　　(**H.** and **C.**)
Ravines and woods ; very common.　June.

376. ASPARAGUS, Linn.

917. **A. OFFICINALIS**, L.　ASPARAGUS.
Slopes of hills and ravines ; frequent on the lake shore from Lake Ridge to Ludlowville Sta.　June.

377. LILIUM, Tourn.

918. * **L. Philadelphicum**, L.　WOOD LILY.　　　　(**H.** and **C.**)
Dry woods ; frequent, especially on the hills.　July.
Fall Creek Woods.　South Hill.　Dart Woods.　Turkey Hill Woods, and elsewhere.

919. **L. Canadense**, L.　　　　　　　　　　　　(**H** and **C.**)
Meadows and swamps ; scarce.　June 20–July.
Beaver Creek and Mud Creek.　Rhodes Woods.　South of Ithaca.

920. * **L. superbum**, L.　TURK'S-CAP LILY.　　　　　　(**C.**)
Meadows and swamps ; frequent.　July.
Case. Woods.　Larch Meadow.　Fall Cr. toward Freeville, etc.

378. ERYTHRONIUM, Linn.

921. * **E. Americanum**, Smith.　YELLOW ADDER-TONGUE　(**H.** and **C.**)
Fields and ravines ; very common.　April–May 15.

379. UVULARIA, Linn.　BELL-WORT.

922. * **U. perfoliata**, L.　　　　　　　　　　　　(**H.** and **C.**)
Along ravines and borders of woods ; frequent.　May.

923. **U. grandiflora**, Smith.　　　　　　　　　　(**H.** and **C.**)
Ravines and rich woods ; common.　May.

380. OAKESIA, Watson.

924. * O. sessilifolia, Wats. (*Uvularia sessilifolia*, L. Man., p. 528.)
(H. and C.)
In ravines and woods; frequent. May.

381. STREPTOPUS, Michx.

925. S. roseus, Michx. TWISTED-STALK. (C.)
Ravines and swamps; abundant. May.

382. PROSARTES, D. Don.

926. P. lanuginosa, Don. PROSARTES. (H. and C.)
Rich woods; infrequent. May.
Six Mile Creek; slope near Beechwoods. McGowan Woods.
Dart Woods. Turkey Hill and Rhodes Woods. Caroline, on Bald
Hill and Taft Hill, and elsewhere.

383. CLINTONIA, Raf.

927. C. borealis, Raf. CLINTONIA. (H. and C.)
Cold swamps or vicinity of peat bogs; frequent. May 25–June.
Ellis Hollow Swamp. Dryden-Lansing Sw., (*Dr. Jordan*)!
Abundant near Freeville, McLean and Danby.

384. MEDEOLA, Gronov.

928. M. Virginica, L. INDIAN CUCUMBER-ROOT. (H. and C.)
Rich woods; abundant. June.

385. TRILLIUM, Linn.

929. * T. erectum, Linn. BIRTHROOT. (H. and C.)
Ravines and rich soil; common. May.
The form called in Gray's Manual "*var. album*" is not uncom-
mon in Six Mile Cr. Negundo Woods. Near Beaver Cr. and Mud
Cr. A form near "*var. declinatum*" of Man. occurs in rich soil of
Negundo Woods and Taughannock ravine; its flowers are usually
pink, rarely nearly white. A form with perianth wholly green oc-
curs in our woods rarely.

930. T. grandiflorum, Salisb. WHITE TRILLIUM. "WHITE LILY,"
(local.) (H. and C.)
Rich soil; very common. May.
Much given to monstrosities. Specimens with four leaves come
from "Camp Warwick," north of King's Ferry; with 4 petals and 4
sepals from Fall Cr., (*F. B. Hine.*) A series of specimens from
woods E. of Levanna show flowers with green-striped petals, and
petals green except on the margin; others with stalked sepals and
green petals, and with long petioled leaves from, or near the base
of the stem. Double-flowered specimens from Woodwardia Swamp
Woods, have about 14 parts to the perianth.

931. T. cernuum, L. NODDING TRILLIUM.
Swamps: rare. May.
Borders of the Round-Marshes! one specimen, 1884, by *F. M.
Fitts*. It ought to occur in Michigan Hollow. It is in Dr. Thomp-
son's Cat., but he probably mistook *T. grandiflorum* for it, as the
latter is not mentioned.

932. * **T. erythrocarpum**, Michx. PAINTED TRILLIUM. (C.)
Cool woods and swamps, where it is frequent. May.
Dryden-Lansing Sw. Ellis Hollow Sw. and frequent in the
swamps of Freeville, McLean, Danby, Newfield, etc.

386. VERATRUM, Tourn.

933. * **V. viride**, Ait. GREEN HELLEBORE. (H. and C.)
Swamps; frequent. July.

387. CHAMÆLIRIUM, Willd.

934. **C. Carolinianum**, Willd. (*C. luteum*, Gray, Man., p. 527.) BLAZ-
ING STAR. (H. and C.)
Woods, especially on the higher hills; sparingly but widely dis-
tributed. June 15-July 15.
Cascadilla and Fall Cr. Woods. Turkey Hill and elsewhere.

388. HEMEROCALLIS, Linn. DAY LILY.

935. H. FULVA, L. (July 1-15,) is established by roadsides. South of
Judd's Falls, in Enfield ravine and in many places.

MUSCARI BOTRYOIDES, Mill., escapes to lawns on E. State street,
Ithaca, etc.; and to roadsides near Taughannock landing, (*F. H.
Severance*.)

PONTEDERIACEÆ.

389. PONTEDERIA, Linn.

936. **P. cordata**, L. PICKEREL WEED.
Marshes, borders of lakes and ponds; common. July 20-Aug.

390. SCHOLLERA, Schreber.

937. **S. graminifolia**, Willd. (*S. graminea*, of Man.) WATER STAR-
GRASS.
Cayuga L., its inlets and outlet; abundant. Aug.

JUNCACEÆ.

391. LUZULA, DC. WOOD-RUSH.

938. **L. pilosa**, Willd. (H. and C.)
Woods and moist pastures; common. April 15-May.

939. **L. campestris**, DC. (H. and C.)
Woods, etc.; frequent. May.

392. JUNCUS, Linn. RUSH.

940. **J. effusus**, L. BULRUSH. (H. and C.)
Wet places; common. July.

941. **J. effusus**, L., var. **conglomeratus**, Engl., occurs south side of
Taughannock, and elsewhere.

942. **J. marginatus**, Rostkov.
Wet pastures; scarce. Aug.
South Hill Marsh. Meadow west of Freeville.

943. **J. bufonius**, L. (H. and C.)
Wet shores and low places; not uncommon. July.

944. **J. Gerardi**, Loisel. BLACK GRASS.
Sparingly, on the brackish marshes E. of Montezuma, north of
main road. July.

945. **J. tenuis**, Willd. (**H. and C.**)
Low grounds and paths ; common. July–Aug.
A slender form, with 3–8 flowers in the panicle and perianth no longer than the pod, occurs on Utt's Point, Cayuga L.

946. **J. tenuis**, Willd., var. ———, tall, (½-¾ meter,) with crowded, glomerate heads ; occurs on the moist meadows by Marl Cr. This is probably the form referred to by Torrey in Flora of N. Y., II, p. 329, as occurring on saline soil of Long Id.

947. **J. articulatus**, L. (**H. and Wright in C.**)
Sandy, wet soil. July. On nearly all the sandy points of Cayuga L. Rarely remote from the lake.

948. **J. articulatus**, L., var. obtusatus, Engl. Depauperate specimens from Trumansburg Point, Utt's Point, and Marl Cr. meadows, apparently belong here.

949. **J. alpinus**, Villars, var. insignis, Fries. July–Aug.
Reduced forms, having the erect stems, light-brown pods, obtuse short inner sepals of that species, and without the fine scalariform wrinklings on the seeds characteristic of all our forms of *J. articulatus* L. are found on Goodwin's (Taughannock) Point, Utt's Point, and by the Indian Salt-Springs west of Cayuga Marshes.[1]

950. **J. acuminatus**, Michx., var. legitimus, Engl. (**H.**)
Wet places ; not uncommon. July.

951. **J. acuminatus**, Michx., var. debilis. "Junius." (*Herb. Sartwell.*)

952. **J. nodosus**, L. (**H.**)
Low grounds ; common. July.
Abnormal forms were collected on Farley's Point (1881) having scales only, in place of essential organs ; and all the scales having undergone chorisis, the heads appear as if made up entirely of trichomes. Where the pistils are only partially transformed, ovules frequently occur along the open scales of the disjoined elements of the ovary.

["**J. scirpoides**, Lam.," is in Sartwell's Herb. from Penn Yan—which is probably a mistake. It occurs at Sodus Bay, (*Dr. Wright.*)]

953. **J. Canadensis**, J. Gay. var. longicaudatus, Engl.
Marshes. Aug.–Sept.
Summit Marsh. Locke Pond. Cayuga Marshes, and Junius Ponds.

954. **J. Canadensis**, J. Gay. var. brachycephalus, Engl.
Wet sandy places ; infrequent. Aug.
Near W. Danby. Dryden Lake. South of Groton. Newton's Ponds, Junius.

TYPHACEÆ.

393. TYPHA, Tourn.

955. **T. latifolia**, L. CATTAIL. (**H. and C.**)
Marshy places, often invading peat-bogs ; common. June.

---

[1] I have recently learned that Engelmann referred specimens like the above, and obtained from Central N. Y., to *J. alpinus, var. insignis.*

956. **T. latifolia**, L. var. **elongata**, n. var.

Leaves very numerous, dark green, elongated, (2-3½ m.) and fruiting spike elongated, often 30 cm. Occurs with the type form in large thick masses near the shore or in the water on Canoga and Cayuga Marshes and north of Hill's Branch. It is the principal form cut for chair-bottoms, and is carried away from Cayuga in considerable quantities for this purpose.

957. **T. angustifolia**, L.　　　　　　　　　　　　　　　(C.)

Abundant on Cayuga and Canoga Marshes; also on west shore of the Inlet near the mouth. Near Etna.

394. SPARGANIUM, Tourn.　BUR-REED.

958. * **S. eurycarpum**, Engln.　　　　　　　　　　　　(H. and C.)

Marshes; frequent. June.

959. **S. simplex**, Huds.　　　　　　　　　　　　　　(H. and C.)

Marshes and standing water; common. July.

960. **S. simplex**, Huds. var. **androcladum**, Gray.

Fall Creek. Summit Marsh.

ARACEÆ.

395. ARISÆMA, Martius.

961. * **A. triphyllum**, Torr.　INDIAN TURNIP.　JACK-IN-THE-PULPIT.
　　　　　　　　　　　　　　　　　　　　　　　　　(H. and C.)

Ravines and wet places; frequent. May.

A specimen from woods east of Freeville has large acuminate leaflets, the lateral with large lobes (7-10 cm. long) on the lower side, thus resembling *A. polymorphum*, Chap., of the South.

962. **A. Dracontium**, Schott.　DRAGON-FLOWER.　GREEN DRAGON.
　　　　　　　　　　　　　　　　　　　　　　　　　(H. and C.)

Rich alluvial soil; scarce. May.

Woods near Fall Cr. west of C. S. railroad. S. W. corner of the lake. Negundo Woods. Woods beyond Larch Meadow, (O. L. Taylor.) Field near the mouth of Paine's Creek. Lockwood's Flats, (*Herb. J. J. Thomas.*)

396. PELTANDRA, Raf.

963. **P. undulata**, Raf.　ARROW-ARUM.　　　　　(H. and C.)

In shallow water; along Cayuga St. on the Marsh. Inlet Marshes. Near Union Springs. Cayuga Marshes. Locke Pond and elsewhere. Our forms, so far as noticed, belong to the above species as restored in Engler's recent monograph. Distinguished from *P. Virginica*, Raf. by the absence of white from the spathe; also the pistillate part of the spadix is ½ or ⅓ the length of the staminate part while in *P. Virginica*, it is ⅔ the length.

397. CALLA, Linn.

964. **C. palustris**, L.　WILD CALLA.　　　　　　(H. and C.)

Low swamps and sphagnum bogs; infrequent. May 20.-June.

Swamp west of the Inlet. Dryden-Lansing Swamp, and Freeville! (*Dr. Jordan.*) Michigan Hollow Swamp and elsewhere.

398. SYMPLOCARPUS, Salisb.

965. *S. fœtidus, Salisb. SKUNK-CABBAGE. (H. and C.)
Marshes ; common. March-April.

399. ACORUS, Linn.

966. * A. Calamus, L. SWEET-FLAG. (H. and C.)
Sedgy marshes ; common. June.

LEMNACEÆ.

400. LEMNA, Linn. DUCKWEED.

967. L. trisulca, L. (H. and C.)
Water, in marshes and ponds ; very abundant about Cayuga Lake.
Jenning's Pond, Danby, Summit marsh and elsewhere.

968. L. minor, L. (H. and C.)
Pools and larger marshes ; abundant.

401. SPIRODELA, Schleiden.

969. S. polyrrhiza, Schleid. (*Lemna polyrrhiza*, L. Man., p. 479.)
(H. and C.)
Pools and marshes ; common.

402. WOLFFIA, Horkel, Schleiden.

970. W. Columbiana, Karst.
Cayuga Marshes, west of Cayuga Bridge. Southwest of the rail-
road bridge over Seneca R. Montezuma Marshes, near the "Marl
Works," and at Mays Point ; usually abundant where it appears.

ALISMACEÆ.

403. ALISMA, Linn.

971. * A. Plantago, L. var. Americanum, Gr. WATER-PLANTAIN.
(H. and C.)
Marshes and borders of streams ; common. July 20-Aug.

404. SAGITTARIA, Linn.

972. * S. variabilis, Englm. ARROW-HEAD. (H. and C.)
All the *varieties* of Man., (excepting var. *pubescens*,) ; common.
July-Aug.

973. S. variabilis, Englm. var. obtusa, Gr., is the best marked variety
and occurs abundantly on the Inlet Marshes and on Cayuga Marshes and elsewhere. In the shallow water of Cayuga L. the broad-
leaved varieties, grow near the shore and var. *hastata*, and var. *angustifolia* farther out. In Cayuga Marshes leaf-blades are found
measuring ¼ m. broad by ½ m. long. In outlet of Summit Marsh
(also at head of Owasco L.) are plants with long phyllodia, 1½-2½
cm. broad, and hastate, floating or erect leaves.

974. S. heterophylla, Pursh. (H. and C.)
Frequent near Cayuga L. Seneca R., and on Summit Marsh.
July-Sept.

975. S. heterophylla, Pursh. var. elliptica, Gr. (also var. *angustifolia*,
Gr.)
Frequent all along the low shores of Cayuga L. and outlet.
[S. graminea, Michx, (*S. acutifolia*, Pursh,) in Dr. Thompson's Cat.
(1841) and Dr. Sartwell's Cat. (1844) as " ?," is probably an error.]

NAIADACEÆ.

405. TRIGLOCHIN, Linn.

976. **T. palustre**, L.

Frequent about Lowery's and Newton's Ponds, W. Junius; "Junius," (*H and C.*)

[**T. maritimum**, L. var. **elatum**, Gr. (*Sartwell, H. and C.*) from Gorham.]

406. SCHEUCHZERIA, Linn.

977. **S. palustris**, L.

Sphagnum Swamp, Junius, (*Sartwell, H. and C.*) June-July.

407. NAIAS, Linn. NAIAD.

978. **N. marina**, L. (*N. major*, of Man., p. 483.)

The typical form is frequent in shallow water near Canoga Marshes and near Cayuga Bridge, west side of lake. July.

979. **N. marina**, L. var. **gracilis**, Morong, (*Bot. Gazette*, vol. 10. p. 254.)

Off Canoga Marshes. Our specimens have the long stems and internodes and long slender leaves of the Florida form from which the original description was drawn; viz. "No. 2705" of Curtiss' "North Amer. Plants." They however lack the purple color of the latter specimens.

980. **N. marina**, L. var. **recurvata**, n. var.

In the shallow water of Black L. Cayuga Marshes. 1880. Stems stoutish, short, 5-12 cm. inclining to be dichotomously branched, spreading and recurved. Leaves purplish, (the olive-purple of *Chondrus*,) recurved usually, the teeth large and prominent. At the base of the leaf the sheath around the fruit shows a projecting tooth on each side, while in the typical form and var. *gracilis*, the sheath is rounded. Seeds, like the type, except reticulations are often more pronounced. Mr. Morong has most kindly examined the *varieties* described by Braun, to which I had not access, and writes that there is nothing corresponding to var. *recurvata* among them.

981. **N. flexilis**, Rostkov. (**H. and C.**)

In all our lakes and quiet ponds. July.

Either mossy and 2-3 cm., or slender and ½ meter or more in length.

408. ZANNICHELLIA, Micheli.

982. **Z. palustris**, L. (**H. and C.**)

Ditches on Ithaca Marshes (*Mr. Lord*)! Occasional in quiet pools along Cayuga L. near King's Ferry, Sheldrake, Canoga and Cayuga Bridge. Near the Indian Salt-Springs west of Cayuga marshes. The so-called var. *pedunculata* is at Cayuga Bridge.

409. POTAMOGETON, Tourn. POND-WEED.

[Dates under this gen us refer to time of fruiting.]

983. **P. natans**, L. (**H. and C.**)

Shallow ponds or stagnant water; common. July-Sept.

984. **P. natans**, L. var. prolixus, Koch. in the Inlet and north of Cayuga Bridge. A form of *P. natans*, with small leaves, resembling *P. Oakesianus*, Robbins, grows in Cayuta L.

985. **P. Claytonii**, Tuck. (**H.**)

Abundant in the still shallow pools of Cayuga and Cayuta Lakes. Aug.-Sept.

Only the specimens from Cayuta L. have the small leaves, as described by Robbins. Cayuga L. specimens usually twice as long, and the petioles usually as long as the blade. The embryo is coiled 1 ½ times, *not* "transversely oval."

986. **P. Spirillus**, Tuck.

Cayuta L. east side, shallow water in sand ; rare. Aug.

987. **? P. rufescens**, Schrad.

Myers Point, and pool two miles north. Summit Marsh ; scarce. In flower in Aug. and Sept. but not found in fruit. Dr. Robbins doubtfully referred them to *P. rufescens* rather than *P. gramineus*.

988. **P. lonchites**, Tuck. (H. and Wright in **C.**)

Fall Cr. near the mouth. Dryden L. North of Cayuga Bridge Erie Canal and Seneca river ; not common. Aug.-Sept.

The Erie canal forms belong to Tuckerman's *P. fluitans*, Roth. described in Silliman's Journal, LVII, p. 348.

989. **P.** (species doubtful : possibly *P. Illinoiensis*, Morong,) was found in the Inlet one season.

990. **P. amplifolius**, Tuck. (H. and Wright in **C.**)

Cayuga and Cayuta Lakes. Dryden L. and Locke Pond. Aug. Good specimens in the bayou at mouth of Fall Cr. Deep water forms are without floating leaves.

991. **P. gramineus**, L. (H. and **C.**)

Cayuga L. near light-house and Myers Pt. Cayuta Lake. Summit Marsh, and elsewhere, extremely variable. Most of our forms correspond most nearly to *var. heterophyllus* of Manual, which cannot, however, be clearly defined as a variety, in this country, according to Mr. Morong.

992. **P. Zizii**, Mertens and Koch. (*P. lucens*, L., *var. minor*, Nolte, of Man., p. 488.) (**C.** by Dr. Wright.)

Fall Creek near the east bayou. Summit Marsh, usually in shallow water ; not rare, but fruits sparingly. Aug-Sept.

993. **P. lucens**, L. (H. and **C.**)

West side of Cayuga L. and near Cayuga Bridge and Seneca River. Aug.-Sept.

994. **P. prælongus**, Wulfen. (**C.** by Dr. Wright.)

S. W. part of Cayuga L. and occasional at the foot of the lake ; not common. Aug.-Sept.

995. **P. perfoliatus**, L. (H. and **C.**)

Very abundant in shallow water from 1-4 meters deep, in Cayuga L., in ponds and creeks. July-Sept.

996. **P. perfoliatus**, L., var. lanceolatus, Robbins.

The leaves vary much in length, but some in Cayuga Outlet should probably be referred to the *var. lanceolatus*.

997. **P. crispus,** L. (**C.** by Dr. Wright.)

Sterile forms exceedingly abundant in deeper water of inlet and lake; fertile in shallower channels and ditches. Rarely fertile near Ithaca in June–July. Abundant in flower and fruit in spring-brook entering west side of Black Lake, Aug. 15, 1885. (Fruits in the Inlet at Watkins.) (Dr. Wright says it has increased in L. Keuka, enormously within a few years, to the exclusion of other species once abundant.)

998. **P. zosteræfolius,** Schum. (*P. compressus*, L., Man., p. 488.)

(**H.** and **C.**)

Frequent in Cayuga L., Summit Marsh, Dryden L., Locke Pond, Mill-ponds at Red Mills and Malloryville. Our forms are robust, resembling *Schollera* in the water.

999. **P. obtusifolius,** M. and K.

Summit Marsh, in the shallow water not far from the northern-most island, where it occurs but sparingly. Aug.

My specimens are short and bushy on account of the habitat, but they otherwise agree closely with specimens from Germany and from Tenn. The latter, sent me by Professor Porter, seem to be nearer the type than his own from "Dillerville Swamp." There are two conspicuous glands at the base of each leaf in all specimens I have seen. These are not noticed in any of the descriptions. Mr. Morong has since written me that they are always present in this rare species.

1000. **P. Hillii,** Morong. (See *Botan. Gazette*, VI, p. 291.)

Cayuga L. on Myers Pt. In pools north and south of Ithaca. In the Inlet. Dryden L. and Red Mills pond; not uncommon. July, Aug.

First collected in 1875 but referred to *P. pauciflorus*, Pursh, until *P. Hillii* was published. It fruits earlier than *P. pauciflorus*. "Stem about 1° long, slender and widely branching. Leaves all submerged, linear, acute 1′–2½′ long, ½″–1¼″ wide, 3-nerved, the lateral nerves delicate and nearer the margins than the midrib, the midrib often compound below. Stipules free, whitish, obtuse, striate, 3″–5″ long. Peduncles short spreading or somewhat re-curved, more or less clavate. Spikes capitate, 3–6 fruited. Fruit obliquely obovate about 1¼″ × 1⅛″, tricarinate on the back, the front slightly arched, obtuse at base. Style nearly facial, short, recurved. Embryo with apex pointing transversely inward." Hitherto reported only from Ohio and westward; but in the *Rob-bins Herb.* from Dr. Hoysradt, of Dutchess Co., (*Mr. Morong.*)

1001. **P. pauciflorus,** Pursh. (**H.** and Wright in **C.**)

Fall Creek near the mouth. Ditches near Esty's Tannery. Near Negundo Woods, (a larger leaved form); not common. Aug.

1002. **P. pusillus,** L., (the *var. vulgaris* of Man., p. 489.) (**H.** and **C.**)

Near mouth of Fall Cr.; scarce and not in fruit.

1003. **P. pusillus,** L., var. **major,** Fries.

Fall Cr. near the mouth, and Cayuga L. west of the breakwater. Cayuga Bridge.

1004. **P. pusillus**, L., var. tenuissimus, M. and K.

Danby, in Jennings Pond; apparently scarce.

1005. **P. marinus**, L.

Cayuga L. near Union Springs. ("Fine specimens" in Seneca L., *Mr. Morong*.) This species resembles *P. pectinatus* but is smaller (usually from 10-20 cm. in length) with slenderer, lighter green leaves and smaller fruits.

1006. **P. pectinatus**, L.   (H. and C.)

Cayuga L. everywhere abundant, and variable. Aug.–Sept.

There are two remarkable forms, perhaps *varieties*, here, viz.:

1007. var. ———? with slender elongated stems (1-1½ meters); nodes remote, as are the whorls of the spike, whose peduncle is usually over ¼ m. long. Leaves few and slender, plants sometimes proliferous. Near the light-house, Cayuga L. Dr. Robbins "found no parallel for this remarkable form," in his own observations.

1008. var. ———? a gigantic form growing in deep water N. W. and N. E. of the light-house Cayuga L. Not yet found in flower or fruit, though examined more or less frequently during ten years past. It is frequently proliferous, especially if detached. It grows in banks, the plume-like bushy tops reaching the surface of the water. The leaves and sheaths are similar to *P. pectinatus*, except in length. Dr. Robbins remarked that he had "nothing that comes near to it in length of leaves—*usque ad* 10'." Stipules are usually much shorter than in *P. pectinatus*. Specimens were obtained in 1874 from 4-5½ meters long. This form was also noticed by Mr. H. B. Lord, probably somewhat earlier than 1874.

1009. **P. Robbinsii**, Oakes.

Abundant in all our lakes, but not yet found in flower or fruit.

CYPERACEÆ.

410. CYPERUS, Linn. GALINGALE.

1010. **C. diandrus**, Torr.   (H. and C.)

Low grounds, and near streams and shores; frequent. Aug.–Sept.

1011. **C. inflexus**, Muhl.

Along the bar in Eddy Pond, Cascadilla Creek (1884); rare. Aug.–Sept.

1012. **C. phymatodes**, Muhl.

Shores, also thriving in damp cultivated ground; infrequent. Aug.–Sept.

South Hill. Linn St., Ithaca. Cayuga L. on the damp sandy points.

1013. **C. strigosus**, L.   (H. and C.)

Low grounds; common and variable. Aug., Sept.

A form is not uncommon on the brackish soil near the "Deerlick" west side of Cayuga marshes, bearing 5-flowered spikelets in narrow oblong rays. (Same form in herb. from Buffalo and Salina, N. Y.)

1014. **C. Michauxianus**, Schultes.　　　　　　　　(H. and C.)

Sandy shores, only toward the foot of Cayuga L. where it is frequent. Aug.-Sept.

Union Springs and Canoga to Cayuga, and on the borders of Cayuga marshes.

1015. **C. Engelmanni**, Steud ; rare, on shore of Canoga marshes. Sept.

1016. **C. filiculmis**, Vahl.

Sandy fields near W. Junius Pout Pond. "Junius," (*H. and C.*) Aug.-Sept.

411. Dulichium, Pers.

1017. **D. spathaceum**, Richard.　　　　　　　　(H. and C.)

Marshes, often on the borders of peat-bogs ; common. July-Aug.

412. Eleocharis, R. Br.

1018. **E. obtusa**, Schultes.　　　　　　　　　　　(C.)

Wet grounds and shores ; common. June-Aug.

1019. **E. ———**, form clearly allied to *E. obtusa*, but with recurved stems, smaller akenes, narrower tubercles, 2-cleft style, and slenderer shorter bristles ; occurs occasionally on low shores of ponds. Eddy Pond. Jenning's Pond, Danby. Dryden L. Stamens not seen, but excepting the habit, this form resembles Ch. Wright's *E. diandra.* (See *Torrey Club Bulletin*, X. p. 101.)

1020. **E. palustris**, R. Br.　　　　　　　　　　(H. and C.)

Marshes and shallows ; common. Over 1 m. tall at Summit Marsh. June-July.

1021. **E. palustris**, R. Br., var. **glaucescens**, Gr., on Canoga Marshes.

1022. **E. rostellata**, Torr.

Rare—July. West side Cayuga Marshes, about the Indian Salt-Springs. Junius, about Newton's and Lowery's Ponds. Probably the latter is Sartwell's old station "Junius," (See *Sartw. Herb. ; Torrey's Fl. N. Y.: Man., 5th Ed.*)

1023. **E. intermedia**, Schultes.　　　　　　　　(H. and C.)

Low shores of lakes, ponds and muddy streams ; frequent. July. Plants often erect when growing in muddy places.

1024. **E. tenuis**, Schultes.　　　　　　　　　　(C.)

Marshes ; frequent ; June.

1025. **E. acicularis**, R. Br.　　　　　　　　　(H. and C.)

Marshes and muddy shores ; common. June-Oct.　　(C.)

1026. **E. pauciflora**, Wats. (*Scirpus pauciflorus, Lightf.*, Man., p .560.)

Rare ; July-Sept. Cayuga L. on Utts Pt. and Farley's Pt. In shallow water, Summit Marsh, scarce, and taller than usual, blossoming about Aug. 15.

413. Scirpus, Linn.

1027. **S. planifolius**, Muhl.

Dry slopes, in woods ; scarce. May-June.

Woods south of Esty Glen, (*F. C. Curtice* and *F. L. Kilborne.*) Renwick Farm slope. South Hill beyond the "Incline." Fall Creek, north of Fiske-McGraw mansion. South side of Paine's Creek, Aurora.

1028. **S. pungens**, Vahl. (**H.** and **C.**)

Sandy shores; frequent near Cayuga L. Locke Pond and elsewhere. July, Sept.

1029. **S. lacustris**, L. (*S. validus*, Vahl., Man., p. 562,) the common
form in E. North Amer. has smaller, narrower akenes than the typical *S. lacustris* which has been found by Mr. O. E. Pearce at Sodus
Bay, N. Y.

Marshes and lakes; common. July–Aug.

Its tall, dark rods are very characteristic of the shallows, forming
a "tule"-like fringe at the north and south ends of Cayuga L.

1030. **S. Smithii**, Gray.

Shores of Cayuga Lake; rare. Aug–Sept.

Myers Pt. Utts Pt and mouth of Big Gully Brook.

1031. **S. maritimus**, L. SEA CLUB-RUSH.

Near salt springs; rare. Aug., Sept.

Montezuma Marshes, (*Dr. Peck, 28th N. Y. Rep.*, p. 89.)! It
occurs by the two salt springs near Salt Cr. (large plants); and in
the brackish shallow water in a pasture east of the Creek, (small
form.)

1032. **S. fluviatilis**, Gray. (**H.** and **C.**)

In the larger marshes, where it is common. June–Aug.

Inlet Marshes. Summit Marshes and elsewhere. One of the
commonest plants on the Cayuga Marshes.

1033. **S. sylvaticus**, L., var. **digynus**, Bœck, (*S. microcarpus*, Presl.,
Man., p. 564.)

Marshy places; not common. June.

Marsh south of the boat-landing and north of Ithaca. Cayuga
Marshes. Near Wood's Sta.

1034. **S. atrovirens**, Muhl. (**H.** and **C.**)

Low grounds; common. June–July.

1035. **S. polyphyllus**, Vahl. (**H.**)

Meadows; scarce. July.

Larch Meadow, White Church valley and near Wilseyville.

1036. **S. lineatus**, Michx.

Wet sandy places; scarce. June 20–July.

By road north of Jacksonville. South of Taughannock ravine.
North of Taughannock ravine, by springs in a field. Near mouth
of Paine's Creek. Genoa, by R. R.

1037. **S. Eriophorum**, Michx. WOOL-GRASS. (**H.** and **C.**)

Low grounds; common. Aug.–Sept.

1038. **S. Eriophorum**, Mx., var. **cyperinus**, Gr.

Howland's Point north of Union Springs, and along Seneca R.

1038ᵃ. The *var. laxus* on Myers Point and elsewhere.

414. ERIOPHORUM, Linn. COTTON-GRASS.

1039. **E. alpinum**, L.

Round Marshes only; very rare. June.

1040. **E. vaginatum**, L.

Sphagnum swamps ; scarce. May 20-June 10.

Larch Meadow. Round Marshes.

1041. **E. Virginicum**, L.

Sphagnum swamps, where it is frequent. July-Aug.

Freeville, Malloryville, Woodwardia Sw., Round Marshes, "Junius," (*Sartwell H. and C.*)

1042. **E. Virginicum**, L., var. album, Gr.

Woodwardia Sw. and W. Junius.

1043. **E. polystachyon**, L.

In nearly all sphagnum swamps. May 25-June 15.

Most of our specimens have rough peduncles, but they are nearly smooth in those at Mud Cr.

1044. **E. gracile**, Koch.   (Dr. Wright in **C.**)

Sphagnum swamps ; occasional. June 20-July 20.

Freeville. S. E. of Chicago Sta. Venice, near R. R. W. Junius Tamarack swamp.

415. RHYNCHOSPORA, Vahl.

1045. **R. alba**, Vahl.

Marshy places and sphagnum bogs ; frequent. July 20-Sept.

1046. **R. capillacea**, Torr.

Rare ; July, Aug., about Newton's Pond. W. Junius. "*Junius*," *Sartw. H. and C.*)

416. CLADIUM, P. Browne.

1047. **C. mariscoides**, Torr.

Marshy places ; scarce. July 20-Oct.

Summit Marsh. Peat-bog S. E. of Chicago Sta. Newton's P., Junius, "*Junius*," (*Sartwell, H. and C.*)

417. SCLERIA, Linn.

[**S. triglomerata**, Michx. in Sartwell's Catalogue, 1844.]

1048. **S. verticillata**, Muhl.

Cold marshes ; July, Aug., about Newton's Pond, W. Junius. "*Junius*," (*Sartw. H. and C.*)

418. CAREX, Linn.

[All dates under this genus refer to the period when *mature perigynia* may be found.]

1049. **C. pauciflora**, Lightfoot.   (**C.**)

Spahgnum bogs ; rare. June.

Round Marsh. "Junius," (*Sartwell, in Herb. Cornell Univ.*)

1050. **C. polytrichoides**, Muhl.   (**H.** and **C.**)

Bogs and springy marshes ; abundant. June.

Specimens from Ringwood and Cayuta L. have bracts, 3-4½ cm. long.

1051. **C. Willdenovii**, Schk.   (**H.** and **C.**)

Not uncommon in woods and on banks of ravines. June 20-July 20.

1052. **C. Steudelii**, Kunth.

Rare ; "Amphitheater," in Six Mile Cr. ravine. June.

1053. **C. bromoides**, Schk. (H. and C.)

Shaded swamps ; abundant. June 1–20.

1054. **C. disticha**, Huds. (*C. Sartwellii*, Dew.)

Rare ; in marsh about Lowery's Pond, W. Junius. ."Junius," (*Sartwell; in Herb. of Ham. Coll.* and *Herb. Cornell U.*)

1055. **C. teretiuscula**, Good. (H. and C.)

Peat-bogs usually ; infrequent. June.

Larch Meadow and the peat-bogs of Freeville, McLean and Locke Pond.

1056. **C. teretiuscula**, Good, var. major, Koch.

With the preceding form, and maturing at nearly the same time.

1057. **C. decomposita**, Muhl. (H. and C.)

"Junius," (*Sartwell in Herb.*) (" Penn Yan," *Sartwell in Herb. Cornell U.*)

1058. **C. vulpinoidea**, Michx.

Low grounds ; very common. July.

1059. **C. stipata**, Muhl.

Low grounds ; common. June 1–20.

1060. **C. alopecoidea**, Tuck. (H. and C.)

Moist open meadows ; scarce. June 15–July 15.

Field S. of E. C. and N. depot. E. of Freeville. Near Marl Cr. Frontenac Id. W. Junius. North of Ludlowville, (*F. C. Curtice.*)

1061. **C. sparganioides**, Muhl. (H.)

Woods and rich slopes ; abundant and variable. June.

1062. **C. sparganioides**, Muhl., var. minor, Boott ; frequent and usually with leaves 5–8 mm. broad. Specimens from Enfield ravine and the Cemetery woods have narrow leaves 3 mm. broad.

[**C. cephaloidea**, Dew., is in Sartwell's Herb. from " Penn Yan," but no true *C. cephaloidea* has been found within our own limits.]

1063. **C. cephalophora**, Muhl. (H. and C.)

Meadows and pastures ; very common. June 20–July.

Mr. Lord found at Ithaca specimens with the upper half of the spikes staminate, and in one case the upper spike entirely staminate. (See 20th N. Y. Rep., p. 409.)

1064. **C. rosea**, Schk. (C.)

Grassy banks and woods ; common. June 15–July 15.

1065. **C. rosea**, Schk., *var. minor*, Boott.; frequent.

1066. **C. rosea**, Schk., *var. radiata*, Dew.; scarce. Six Mile Creek " Narrows."

1067. **C. retroflexa**, Muhl. (H. and C.)

Slopes, more or less rocky ; not common. June 10–30.

Casc. Cr. N. W. of Eddy dam. South Hill. Renwick Farm slope and elsewhere.

1068. **C. chordorhiza**, Ehrh.

Cold sphagnum swamps ; "Junius," (*Sartwell in C. U. Herb.*, also in *Herb. Ham. Coll.* and *Cat. of 1844.*)

1069. **C. tenella**, Schk.   (**H.** and **C.**)

Sparingly in the moss of nearly all our cold shaded swamps, especially of Hemlock or Tamarack ; also South Hill Marsh.   June.

1070. **C. trisperma**, Dew.

Swampy woods and mossy swamps ; not uncommon.   June.

1071. **C. canescens**, L.   (**H.** and **C.**)

Cold swamps or peat-bogs ; not uncommon.   June.

1072. **C. canescens**, L. var. vitilis, Carey.   (**H.** and **C.**)

Infrequent.   Dryden-Lansing Sw.   Michigan Hollow Sw.   Freeville bog.   Round Marsh woods, and Enfield Spruce Sw.

1073. **C. Deweyana**, Schk.   (**H.** and **C.**)

Woods.   Six Mile Cr.   McGowan Woods and elsewhere ; not common.   June-July 15.

1074. **C. sterilis**, Willd.   (**H.** and **C.**)

Sphagnum swamps ; scarce.   June 20-July.

Specimens mostly barren, from Round Marsh, Locke Pond and W. Junius ponds ; with fuller fruit at Freeville, Summit Marsh and Locke Pond.

1075. **C. echinata**, Murr.   (*C. stellulata*, Man., p. 579.)

                                                        (**H.** and **C.**)

Open bogs or moist grassy places ; abundant.   June.

1076. **C. echinata**, Murr., var. microcarpa, Boeck.   (*C. stellata var. scirpoides*, Carey.)

In sphagnum bogs ; common.   June.

1077. **C. scoparia**, Schk.   (**H.** and **C.**)

Meadows and wet pastures ; very common.   June 20-July.

1078. **C. scoparia**, Schk. var. intermedia, Olney,—a tall form with oblong aggregated spikes occurs in Fleming Meadow and Summit Marsh.

1079. **C. lagopodioides**, Schk. (typical form of Schkur's fig. 177 and Boott's Ill. tab. 370.)   (**C.**)

Common in low meadows and borders of pools.   July.

1080. **C. lagopodioides**, Schk. var. moniliformis, Olney.

In cold shaded swamps or borders of swamps ; not common. June-July.

1081. **C. cristata**, Schw.   (**H.** and **C.**)

Marshes and along streams : common.   July-Aug.

1082. **C. cristata**, Schw. var. mirabilis, Boott.   (**H.** and **C.**)

Fields and wet meadows ; frequent.

1083. **C. adusta**, Boott.

Woods rare, in the woods on Taft's Hill, Caroline.   July.

1084. **C. straminea**, Schk. (*typica.*)   (**H.** and **C.**)

Moist, grassy places.   South Hill.   Taughannock.   Frontenac Id. and elsewhere. (E. of Watkins.)   June 10-25.

1085. **C. straminea**, Schk. var. tenera, Carey.

Moist thickets ; not common.   June 15-30.

South Hill with *Andromeda*. Near Lucifer Fall, Enfield. Olney's *forma erecta* of this species occurs near Enfield Spruce Swamp.

1086. **C. straminea**, Schk. var. **festucacea**, Carey.

Canoga, the beach near the Marshes. (In Herb. from *Yates Co., Dr. Wright.*)

1087. **C. alata**, Torr.

The larger marshes; scarce. June 15-30.

Northern part of Cayuga Marshes. Montezuma Marshes, east side of Savannah "island." In Herb. from Junius, (*Wright.*) (In C. U. Herb. from Penn Yan, *Sartwell.*)

1088. **C. aquatilis**, Wahl.   (C.)

Cold marshes; not common. June 20-July 10.

Larch Meadow. Round Marshes. Union Springs. Newton's P., Junius. A form with slender leaves and short spikes, at Locke Pond, corresponding to the *C. xerocarpa* form of *C. angustata*.

1089. **C. torta**, Boott.   (H. and C.)

Along ravine beds, and wet places; frequent. May 10-30.

1090. **C. angustata**, Boott. (*C. stricta*, Man., p. 583; *Boott, Illustr.* tab. 586.)   (H. and C.)

In tussocks, in standing water; not common. June 1-20.

Not near Ithaca, but in Round Marsh; and by R. R. west of Malloryville. Hemlock Cr., Groton. Locke Pond.

1091. **C. angustata**, Boott. var. *β*. (*Boott, Illustr.* tab. 587.)

Freeville, mouth of Mud Cr. W. Danby.

1092. **C. angustata**, Boott. var. strictior. (*Boott, Illustr.* tab. 588.)

Frequent; often in broad beds, or in tussocks when in water. Fall Cr., Cold Spring Marsh. Indian Spring Marsh. Round Marshes. Michigan Hollow. Lowery's Pond, Junius; "Junius," (*Sartwell.*)

1093. **C. angustata**, Boott. var. xerocarpa, L. H. B., is at Freeville.

1094. **C. crinita**, Lam.   (H. and C.)

Wet places, common; occasionally androgynous. June 10-30.

1095. **C. gynandra**, Schw. (*C. crinita*, var. *gynandra*, Schw. and Torr.)

Rare; Freeville near sphagnum bog and Fir-Tree Swamp.

1096. **C. limosa**, L.

Sphagnum swamps; scarce. May 20-July.

Freeville bog. Malloryville. Junius, (*Sartwell, H. and C.*); north of the Pout Pond.

1097. **C. Magellanica**, Lam. (*C. irrigua*, Sm. Man., p. 584.)   (C.)

Rare; Freeville Bog. Lowery's Pond, Junius.

1098. **C. Buxbaumii**, Wahl.

Rare; "Junius," (*Sartwell, Herb. Cornell Univ.* and *Herb. Ham. Coll.*)

1099. **C. aurea**, Nutt.   (H. and C.)

Wet springy ground; infrequent. June–July.

Casc. Woods. Six Mile Cr. and elsewhere. Specimens with androgynous terminal spikes occur at Otter Creek, Cortland; in Casc.

Woods and at Taughannock. Spikes compound (proliferous from lower perigynia) at Taughannock, (*F. V. Coville.*)!

1100. **C. tetanica**, Schk.

Moist woods and ravines ; infrequent. June.

Case. Woods, Green-Tree Falls. Freeville, near Fir Swamp, Danby. Round Marsh. Near Woods Sta., with *Jeffersonia*. Junius, Lowery's Pond. "Junius," (*Sartwell C. U. Herb.*)

1101. **C. granularis**, Muhl.   (**H. and C.**)

Grassy meadows and wet places ; common. June 10-20.

Very variable. A broad-leaved form suggests *G. glaucodea*. The "var. *recta*" Dew., with smaller ovate perigynia is the most abundant. Specimens from Case. Woods show an androgynous upper spike. It is the most abundant sedge on the rich farms of Venice.

1102. **C. pallescens**, L.   (**H. and C.**)

In pastures of cold clay soil ; not common. June 15 July.

The *C. undulata* form, (*Kunze*, tab. 4, f. 2.) occurs near Malloryville and Turkey Hill. The Malloryville specimens also had cylindrical proliferous pistillate spikes, and peduncle of staminate spike 2-3 cm. long.

1103. **C. grisea**, Wahl.   (Dr. Wright in **C.**)

Rich shaded soil ; scarce. June.

Case. ravine, (*O. E. Pearce.*) Fall Cr., a few. South Hill Marsh. Near Fall Cr., at Freeville. Along Marl Cr.

1104. **C. glaucodea**, Tuck. (*Proc. Amer. Acad.* VII. p. 395.-1865.)

Rare ; South Hill, with *Adromeda*, 1882. (*F. C. Curtice.*) June 15-30.

Also about South Hill Marsh.

"Spikes 4-5, cylindrical; terminal staminate, subsessile, clavate ; the others pistillate, leafy-bracted, many-flowered, upper approximate, peduncles exserted, lower distantly remote, all with the leaves pale or glaucous-green. Perigynia ovoid-turgid, obtuse, orifice subentire, many-nerved, $1\frac{1}{2}''$ long. Scales broad-ovate, short-cuspidate, whitish, barely 3-nerved and green to the middle. Akenes obovate, triquetrous." Plants usually *spreading or ascending 6'-12' high, smooth. Leaves* $2\frac{1}{2}''-4''$ *wide*, shorter than culm, *very glaucous. Perigynia more dense than in C. grisea, fewer-nerved, pointless, its scale shorter and less pointed.* Occurs sparingly from Mass. to Del. In N. Y. reported only from Dutchess Co. In habit, texture, character of scales and color it seems quite distinct from *C. grisea.*

[**C. conoidea**, Schk. from Penn Yan, *Sartwell, Herb. Cornell U.; also in Cat.*]

1105. **C. gracillima**, Schw.   (**H. and C.**)

Woods ; abundant. June.

Malloryville specimens exhibit proliferous branches to spikes ; and occasionally specimens exhibit smaller, sharper perigynia than usual.

[**C. formosa**, Dew. "Penn Yan." (*Sartwell, H. and C.*]

1106. **C. triceps**, Michx.   (**H. and C.**)

Grassy copses; rare near South Hill Marsh. It varies into the form "*C. hirsuta*, var. *pedunculata*," S. and T., which has oblong cylindrical spikes and leaves scarcely pubescent. Near Bear Swamp.

1107. **C. virescens**, Muhl. (H. and C.)

Woods; infrequent. June 20-July.

Enfield ravine. Near Danby. W. shore of Cayuga L., near Marion's. Near Etna. Freeville.

1108. **C. virescens**, Muhl. var. **elliptica**, Olney. (Spikes longer and perigynia more pointed) is found near Etna. Ovid. Summit Marsh.

1109. **C. plantaginea**, Lam. (H. and C.)

Ravines and rich woods, where it is frequent. May 10-30.

Six mile Creek. Round-Marsh Woods and elsewhere.

1110. **C. Careyana**, Torr.

Rare; May 25-June 10. Taughannock ravine? 1874, (*R. Yatabe*.) Ravine north of Buttermilk Cr., 1882, (*F. C. Curtice*)! Woodwardia Swamp Woods, 1883, (*O. E. Pearce*)! Wood's Sta., with *Jeffersonia*. The type specimens were found near Auburn, N. Y., by John Carey, 1832.

1111. **C. platyphylla**, Carey. (Dr. Wright in C.)

Ravines and beech woods; common. June 1-15.

1112. **C. platyphylla**, Carey, var. ———.

In the nook, opposite the Beechwoods, Six Mile Cr., 1883, (*O. E. Pearce.*) Elmwood Glen, Cayuga L. It differs from the type in leaves broader, (3 cm.) and longer than the culm; and pistillate spikes mostly with 6 flowers instead of 3 or 4; Boott says there are rarely 7 flowers in the type.

1113. **C. retrocurva**, Dew.

Beech woods and elsewhere; abundant. June 15-July 15.

Variable in width of leaves and length of bracts. Forms with green leaves 4-6 mm. broad, in Six Mile Cr., Taughannock and Paine's Cr. These have, however, the sterile flowers at the base of the pistillate spikes, characteristic of true *C. retrocurva*. A form was found near Ovid, with broad leaves, (8-14 mm.) and large perigynia, suggesting *C. platyphylla*, but the peduncles were long and slender.

1114. **C. digitalis**, Willd. (H. and C.)

Woods along ravines and hillsides; common. June 15-July 15.

1115. **C. laxiflora**, Lam. (H. and C.)

Woods, ravines, or even open meadows; common.

The typical form is abundant and earlier than *var. blanda*. Specimens from Green-Tree Falls are proliferous from the lower perigynia.

1116. **C. laxiflora**, Lam., var. **plantaginea**, Boott.

Scarce; genuine specimens from the ravines north of King's Ferry. Less pronounced ones from Casc. Cr. Fall Cr. in Cold Spring Marsh. McGowan Woods and Freeville.

1117. **C. laxiflora**, Lam., var. **intermedia**, Boott.
    Frequent; a peculiar green-leaved, swamp form is referred to this; the latter matures in July.

1118. **C. laxiflora**, Lam., var. **blanda**, Dew.
    Common; the sub-variety *minor* is frequent in the meadows.

1119. **C. laxiflora**, Lam., var. **latifolia**, Boott.
    Rich soil, ravines or old woods; frequent. Six Mile Cr. Enfield ravine. Woods near Freeville. Round Marsh, and elsewhere.

1120. **C. Hitchcockiana**, Dew.   (C.)
    Low woods; scarce. June 15–30.
    North of Forest Home, (*F. C. Curtice.*) South of Danby village. Wood's Sta. Ovid woods. Franklin's ravine and "Camp Warwick."

1121. **C. oligocarpa**, Schk.   (H. and C.)
    Shaly banks of Six Mile Cr., below Green-Tree Falls, 1884, (*O. F. Pearce.*)! Rare; May 25–June 10.

1122. **C. eburnea**, Boott.   (H. and C.)
    Cliffs, or marly places near ravines; not rare. May 15–June 10. Casc. Cr. Fall Cr. Salmon Cr., and elsewhere.

1123. **C. pedunculata**, Muhl.   (C.)
    Shaded ravines, and old logs in swamps; frequent. May 15–30. Cascadilla and other ravines. Mud Cr. and Beaver Cr. swamps.

1124. **C. umbellata**, Schk.   (H. and C.)
    Banks of ravines, and dry slopes; not uncommon. May.
    This is our earliest-flowering Carex.

1125. **C. Emmonsii**, Dew.
    Woods, ravines and hillsides; frequent. May 20–June 20.

1125ª. **C. Emmonsii**, Dew., var. **elliptica**, Boott, occurs south of Shurgur's Glen.

1126. **C. Pennsylvanica**, Lam.   (H. and C.)
    Woods and dry banks; common and variable. June.
    A form is found on Ball Hill in woods, with long culms, nearly ½ m. tall, and leaves nearly ¾ m., much exceeding the culm, as do the slender bracts.

1127. **C. varia**, Muhl.   (C.)
    Woods and rich slopes of ravines; common. June.
    The spike in this varies in size according to soil.

1128. **C. pubescens**, Muhl.   (H. and C.)
    Rich soil; not common. June–July 15.
    Near the Cold Spring. Six Mile Cr. Freeville. Taughannock. McGowan Woods. Enfield. Paine's Cr. and elsewhere.

1129. **C. miliacea**, Muhl.   (H. and C.)
    Wet places in woods; frequent. Fall Cr., etc. May 20–June 20.

1130. **C. scabrata**, Schw.   (H. and C.)
    Cold marshes in ravines and swamps; frequent. July–Aug.

1131. **C. arctata**, Boott.   (C.)
    In rich swampy woods; not common. June.

1132. **C. debilis**, Michx. (C.)

Sunny banks, on the borders of swamps ; scarce. June.

var. a. Boott. (Staminate spike usually not androgynous ; pistillate nodding and broader,) occurs near Dryden-Lansing Swamp, Marl Cr., Marl Pond woods, Bear Swamp and near the Pout Pond, W. Junius.

1133. *var. β*, Boott, (staminate spike with some pistillate flowers ; pistillate spikes slender, erect,) is found near Freeville bog, in the meadows southeast of Etna and on Turkey Hill.

1134. **C. capillaris**, L., var. elongata, Torr.

On half-submerged logs west side of the principal "Marl Pond," South Cortland, 1884. ("Otter Cr. near Cortland, N. Y.; *S. N. Cowles*, 1869 ;" see *Olney's Carices, Boreali-Amer.*) At the latter station,—3 miles from the former,—many of the plants were destroyed in constructing the fish-pond ; but others still exist a few rods north, on the logs which lie in the cold, spring-water. Not elsewhere known in N. Y., although it probably exists within its limits. The home of the species is in the Rocky Mt. region from Col. to Alaska. Said to occur in N. H.

1135. **C. flava**, L.

Spring meadows and marshes, where it is common. June–July.

The form, "*var. androgyna*," Olney, is at Summit Marsh, Locke Pond, the Marl Ponds, and at the mouth of Paine's Cr. Some of the Locke Pond specimens possess compound pistillate spikes. (Junius, *H. and C.*)

1136. **C. Œderi**, Ehrh.

Limestone or marly soil ; scarce but variable. June–Sept.

The form nearest to the type,—yet of a very different aspect, is the *var. prolifera*, H. B. Lord, *19th N. Y. Rep.*, 1865, p. 76, in which "at least one of the spikes of each plant is proliferous ; that is, from one of the perigynia issues a stalk bearing a spike" ; usually the secondary spikes are abundant enough to form a dense ovoid mass or irregular head. It is altogether probable that Michaux, (*Flora Bor. Amer.*, II, p. 171,) had similar specimens in mind in describing the spikes of *C. Œderi* as "many, conglomerate-approximate." Found near Cayuga L.; at Salmon Cr., south of the spring; (*Mr. Lord,*)! Farley's Point ; Utts Point ; shore near quarry south of Union Springs.

1137. **C. Œderi**, Ehrh., *forma intermedia.*

Near the Marl Ponds and Marl Cr. meadows. Near Otter Cr. These are nearest the slender forms, (*C. viridula*, Michx., formerly,) of *C. Œderi*, such as those at W. Bergen, in the great marl-swamp ; but are still taller and more robust, approaching in habit *C. flava.* They have the dark and rather slender rigid, erect leaves and bracts, and the small perigynia mostly ascending with a short straight beak, characteristic of *C. Œderi* ; but the spikes are oblong or cylindrical, mostly 1½ cm., robust, the upper variously androgynous, the lower often proliferously compound.

1138. **C. filiformis**, L..   (**C.**)

Sphagnum swamps, where it is frequent. June.

1139. **C. filiformis**, L., var. latifolia, Boeck., (*C. lanuginosa*, Man., p. 595.)   (**H.** and **C.**)

Borders of most of our sphagnum marshes; also near the shore on Farley's Point and in marsh north of Union Springs.

1140. **C. riparia**, Curtis.   (**H.** and **C.**)

Sedgy sphagnum swamps, where it is frequent. June 15-July.

It also occurs in springy soil on the Inlet and other marshes.

1141. C. HIRTA, L.

Introduced by D. L. and W. railroad, on South Hill east of the "Incline," where it is spreading vigorously. Detected 1882, but may have been growing there for ten years. It is a native of Europe and Asia. Elsewhere introduced in Amer. only at Philadelphia, and in Mass. Among our native species, its closest affinity seems to be with *C. trichocarpa*, with which it agrees in its running rootstocks, hairy perigynia, and its scale. It is a much lower plant than that species, its leaves and sheaths are soft-hairy, and the pistillate spikes remote and scattered. The length of the perigynium is 6-7 mm., and the scale is aristate.

1142. **C. trichocarpa**, Muhl.   (**H.** and **C.**)

Along the larger marshes or lakes; infrequent. June 20-July.

Fall Cr., Beebe Pond. Marshes about Ithaca. Ludlowville. North of Freeville, and elsewhere.

[**C. trichocarpa**, Muhl., var. imberbis, Carey, at Penn Yan, *Gray's Man.*, p. 597.]

1143. **C. comosa**, Boott.   (**H.** and **C.**)

Borders of ponds; not common. July.

Marshes along Cayuga St. Round Marsh Pond. Cayuta L. and elsewhere.

1144. **C. comosa**, Boott. × **C. tentaculata**, Muhl., was found near W. Danby.

1145. **C. Pseudo-Cyperus**, L. × **C. hystricina**, Willd. ?

Near the Pout Pond, W. Junius.

Both of these apparent hybrids present intermediate characteristics between the supposed parent species, and both have well-developed but empty perigynia. In the case of the former, (No. 1144,) the two parents were growing in the closest proximity.

1146. **C. Pseudo-Cyperus**, L.   (**H.** and **C.**)

Rare; at Locke Pond; (and by the Erie Canal, east of the Lockpit.) Wm. Boott's view, viz: that *C. comosa* should be reduced to a *variety* of this species, is no doubt the correct one.

1147. **C. hystricina**, Willd.   (**H.** and **C.**)

Open meadows; frequent. June 15-30.

Along Cayuga L., the upper spike is occasionally androgynous, and all the spikes more or less compound.

1148. **C. tentaculata**, Muhl.   (**H.** and **C.**)

Meadows and swamps; common. July-Aug.

1149. **C. intumescens,** Rudge. (H. and C.)

Chiefly near the colder swamps ; frequent. June–Aug.

South Hill Marsh, north of Forest Home, and elsewhere.

1150. **C. Grayii,** Carey.

Shaded swamps ; rare. June–July.

Freeville, south of Fir-Tree Swamp, 1882, (*F. C. C.* and *W. R. D.*) (Head of Owasco L., with *Carya Sulcata,* 1881.)

1151. **C. lupulina,** Muhl. (H. and C.)

Swamps ; common and variable. June 20–Aug.

Beside the typical form, having the fertile spikes approximate, there are several well-marked conditions, as follows : 1. A form with stalked and scattered fertile spikes, frequent and approaching in aspect *C. Halei, Boott, (Illustr.* II, t. 279.) 2. The "*var. longi-pedunculata,* Sartwell," (see *Horace Mann Herb. C. U.*) having 4–7 spikes, the pistillate on long peduncles, the longest 20–25 cm. ; scarce near Ellis Hollow Swamp. 3. A large robust form ; leaves 10–14 cm. broad, spikes and perigynia large and akenes broader than usual but mostly abortive ; Woodwardia Swamp and Myers Point. 4. The well-marked *var. polystachya,* Schw. and Torr., (*C. lupuliformis,* Man., p. 598,) in *Cornell U. Herb.* and *Herb. of Ham. Coll.,* from Penn Yan, (*Sartwell.*)

1152. **C. lupulina,** Muhl. × **C. retrorsa,** Schw., (*C. lupulina,* Muhl., *var. gigantoidea,* Dew., *Amer. Jour. of Arts and Sci.,* XCI, 1866, p. 328.) The specimens from which Dewey described his *var. gi-gantoidea* were collected by Hon. H. B. Lord, on Myers Pt. in 1865 ; and his description, and the specimens still to be found there, agree substantially with those found rather frequently on the Ithaca Marsh between the Glass Works and Willow Ave. ; north of Free-ville ; and near Taughannock Sta. of C. S. railroad, (*F. C. C.*) After carefully examining about a hundred specimens, there seems little doubt that they are hybrids of the above, for the following reasons : 1. The diverging perigynia are intermediate in size and number of nerves. 2. The 2–3 pistillate spikes are scattered as in *C. lupu-lina,* (form 1,) and mature between the time of ripening of *C. retrorsa* and of *C. lupulina.* 3. There are usually 2–3 staminate spikes with the aspect of *C. retrorsa,* (*C. lupulina* has but one,) often bearing a few perigynia at the base from which small staminate spikes are frequently proliferous. The latter condition is often seen in *C. re-trorsa* and is shown in Boott's tab. 276 and described on p. 94, Vol. II. 4. Akenes almost always abortive, as is usual with hybrids ; when present not well-formed. 5. When mature the perigynia re-semble the inflated ones of *C. lupulina* in texture and appearance, but are straw-colored like *C. retrorsa.* 6. The hybrids are usually accompanied by both the above species. It is interesting to find this form in the Herb. of Prof'r Thomas, collected in 1827.

1153. **C. folliculata,** L. (H.)

Low woods ; scarce. June–Aug.

South Hill Marsh. North of Forest Home in Woods. Freeville, in Fir-Tree Swamp. North of Etna. Beaver Cr. Swamp.

[C. squarrosa, L., in Sartwell's Cat of 1844.]

1154. **C. retrorsa**, Schw. (**H. and C.**)

　　Meadows and the lake shore; frequent. June-July.

1155. **C. retrorsa**, Schw., var. **Hartii**., Carey, (for full description, see Paine's Cat., p. 163.) Ludlowville, (*Mr. Lord* in Paine's Cat.) Ithaca Marsh. Ellis Hollow Swamp. Slenderer than the type, the lowest spikes long-peduncled, perigynia few-nerved and akenes not as papillose as in the type.

1156. **C. utriculata**, Boott. (**C.**)

　　The typical form, in the borders of deep cold swamps, is rare here; Summit Marsh. Meadows south of Michigan Hollow Swamp, as *forma composita* (local.) These have elliptical or oblong perigynia, 7-9 mm. long, with short teeth, as figured in *Boott, Illustr.*, tab. 39. The spikes of the latter are branched below in a proliferous manner. The common form, occurring in the larger marshes and along our ponds and lakes, has much smaller, nearly ovate perigynia. Its extreme is the next variety.

1157. **C. utriculata**, Boott. var. **minor**, Sartwell.

　　Ludlowville, (*H. P. Lord, in Paine's Cat.*, p. 164, and in *Cornell Univ. Herb.*) Canoga. Round Marsh.

1158. **C. ampullacea**, Good. (**H.**)

　　Specimens from Locke Pond, and more pronounced ones from the Marl Ponds, must be referred here, although they are not typical. The few spikes, slender and mosly smooth culm, canaliculate leaves, and scarcely hispid scale, show the approximation to that species, and yet suggest certain characters of *C. utriculata*.

1158². **C. ampullacea**, Good., var. **sparsiflora**, Dew., "Junius," (*Herb. Sartwell.*)

1159. **C. monile**, Tuck. (**H and C.**)

　　Borders of pools, etc.; rare. July.

　　Near the road south from the Junius Pout Pond. (Near Moravia, *Dr. Atwood.* Yates Co., *Cornell U. Herb.*, from *Sartwell and Wright.*)

1160. **C. Tuckermani**, Boott. (**H.**)

　　Marshes and low woods; infrequent. June-July.

　　Marsh north of Ithaca. Ludlowville, (*Mr. Lord.*) Ringwood. By Fall Cr. near Freeville, and elsewhere.

[C. longirostris, Torr. Rare E. side of the valley at Watkins. June 1-12. Penn Yan, *Sartwell, in Herb. C. U.*]

GRAMINEÆ.

419. PANICUM, Linn.

1161. **P. GLABRUM**, Gaudin. (**H. and C.**)

　　Lawns and roadsides; frequent near Ithaca. Aug.-Sept.

1162. **P. SANGUINALE**, L. CRAB-GRASS. (**H and C.**)

　　Cult. soils; everywhere. July-Oct.

1163. **P. proliferum**, Lam.

　　Brackish soils; local. Aug.-Sept.

　　Fields near the Salt Creek, Montezuma.

1164. **P. capillare,** L. OLD-WITCH GRASS.

Cult. soils ; common. Aug.–Oct.

A form with spindle-shaped spikelets and all parts slenderer than type is frequent on the lake shore.

1165. **P. virgatum,** L.

Near Cayuga L. ; rare. Aug.

Farley's Point, a single cluster, 1881. "Cayuga Marshes," (*Sart-well, in Paine's Cat.,* p. 173.)

1166. **P. latifolium,** Muhl. (H. and C.)

Woods ; frequent. July.

Beside the typical form, a slender one is frequent, usually branched and destitute of hairs at the joint, leaves elongated-lanceolate. It is abundant in Casc. Woods, and elsewhere.

1167. **P. clandestinum,** L. (H. and C.)

Rare ; only in the border of the woods near the brook north of the "Nook." Possibly introduced here.

1168. **P. microcarpum,** Muhl., var. **sphærocarpon,** Vasey.

Rare ; woods near the road between the W. Junius ponds and Geneva.

1169. **P. nervosum,** Muhl. (*P. commutatum,* Schultz. *Vasey, Cat. of* 1885, p. 9.)

Woods : not rare. July.

Resembles broad-leaved forms of *P. dichotomum,* L. Culms are 30–40 cm. ; smooth ; leaves elongated, lanceolate-acuminate, 8–15 cm. long, 1–1½ cm. wide ; scabrous near the margins, and ciliate with long hairs near the base and along the sheaths ; panicle rather ample ; spikelets 2 mm. long, slightly hairy. Casc. Woods. Fall Creek Woods. Near South Hill Marsh. Near White Church. Coy Glen, (*Prof. W. Trelease*) !

1170. **P. xanthophysum,** Gray.

High hills in dry woods ; rare. July 1–25.

Summit of the South Pinnacle, Caroline. Thacher's Pinnacle, W. Danby. (Near Painted Post, *Miss Arnold.*)

1171. **P. dichotomum,** L. (H. and C.)

Woods and dry banks ; common and variable. July, Aug.

Among the less hairy forms belonging to the type is the provisional var. *viride,* Vasey, growing in woods slender, with narrow leaves which spread from the culm at right angles.

1172. **P. dichotomum,** L. var. **nitidum,** Gr.

Casc. and Fall Creek Woods, and elsewhere ; frequent.

1173. **P. dichotomum,** L. var. **pubescens,** Vasey, (*P. pubescens* ? Man. 649,) may at present stand for a large number of more or less hairy forms frequent in this region. The hairs in some, extend up the peduncle into the panicle ; all have hairy flowers. This species needs a thorough revision.

1174. **P. depauperatum,** Muhl. (H. and C.)

Woods and banks ; frequent. June.

1175. P. CRUS-GALLI, L.   BARN-YARD GRASS.   (H. and C.)
Cultivated grounds, shores, etc. ; very common.   Aug.-Sept.
Beside the common form it varies into the following :

1176. P. CRUS-GALLI, L., var. MUTICUM, Vasey.
The flowers are almost wholly awnless, the clusters close and
dark, often blackish-purple.   Myers Point.   Near Union Springs,
Cayuga and Montezuma.

1177. **P. Crus-galli**, L., var. hispidum, Gray, (clearly the *P. hispidum.*
Muhl., *Descript. Uberior Gram.* p. 107 ; probably *P. echinatum*,
Trin., but not clearly that, ; see fig. 162, in *Species Graminum.*)
Inlet Marshes.   Cayuga Marshes, where its large violet-colored
panicles are very showy.   Probably indigenous.

P. MILIACEUM, L.
Occasionally spontaneous.   By the old roads on the marsh north
of Ithaca.   Aug.

420. SETARIA, Beauv.   FOXTAIL.

1178. S. GLAUCA, Beauv.   (H. and C.)
Fields ; common.   July-Sept.

1179. S. VIRIDIS, Beauv.   (H. and C.)
Fields ; common.   July-Sept.

1180. S. VIRIDIS, Beauv., var. PURPURASCENS, Peck.
Spike narrower, bristles few and purplish.   The spike and flow-
ers of our form are still smaller than Dr. Peck's type specimens.
Near Union Springs.

421. SPARTINA, Schreb.

1181. **S. cynosuroides**, Willd.   CORDGRASS.
Near Cayuga L.   Aug, Sept.
Farley's Point, a few.   Dwarf but rather abundant on Cayuga
Marshes, north of Black Lake.

422. ZIZANIA, Linn.

1182. **Z. aquatica**, L.   WILD RICE.   (H. and C.)
The larger marshes, where it is frequent.   July 25-Aug. 15.
Inlet marshes.   Cayuga Marshes.

423. LEERSIA, Swartz.

1183. **L. Virginica**, Willd.   (H. and C.)
Shaded places ; not uncommon.   Aug.-Sept.
Near the Indian Spring.   Negundo Woods.   Near Fall Cr., Free-
ville and elsewhere.

1184. **L. oryzoides**, Swartz.   CUT GRASS.   (H. and C.)
Wet sedgy ground ; common.   Aug.-Sept.

424. ANDROPOGON, Linn.

1185. **A. provincialis**, Lam. (*A. furcatus*, Muhl.)   (H. and C.)
Dry gravelly banks ; frequent.   Aug.-Sept.
Fall Cr.   Case. Cr. and elsewhere.   Especially abund. along the
lake-shore.

1186. **A. scoparius**, Michx.   (H. and C.)
Dry soil ; common.   Aug.-Sept.

[**A.** dissitiflorus, Michx. (*A. Virginicus*, L.,; Penn Yan, *Sartwell*, *Herb. and C.*]

425. CHRYSOPOGON, Trin.

1187. **C.** nutans, Benth. (*Sorghum nutans*, Man., p. 652. (**H.** and **C.**)
Dry banks ; not uncommon. Aug. 15–Sept. 15.
Fall Cr., near Triphammer Falls. Six Mile Cr. Lake shore and elsewhere.

426. PHALARIS, Linn.

P. CANARIENSIS, L. CANARY GRASS.
Appears nearly every year on the campus and in waste places near Ithaca. Not established. July.

1188. **P.** arundinacea, L. (**H.** and **C.**)
Marshes, where it is frequent. June.
Especially on Inlet and Cayuga Marshes.

1189. * P. ARUNDINACEA, L. var. PICTA.
Established by road near steamboat landing, Ithaca. East of Varna, by railroad. Near Red-Mills Pond. Near Dryden L.

427. ANTHOXANTHUM, Linn.

1190. A. ODORATUM, L. SWEET-SCENTED GRASS.
Fields ; infrequent. June.
Univ. campus. By road south of Beebe Pond. W. Dryden. Caroline Hills. Saxon Hill.

428. ALOPECURUS, Linn.

1191. A. PRATENSIS, L.
Scarce ; a few places on the campus. May 20–30.

1192. A. GENICULATUS, L. (**C.**)
Scarce; Canoga Marshes. Cortland Marl Ponds.

1193. **A.** geniculatus, L. var. aristulatus, Munro. (*A. aristulatus*, Mx. Man., p. 608.) (**H.** and **C.**)
Marshes and near peat-bogs ; frequent. June.
Fall Creek and Inlet Marshes. South Hill Marsh. Dryden-Lansing Swamp. Summit Marsh and elsewhere.

429. ORYZOPSIS, Michx.

1194. **O.** melanocarpa. Muhl. (**H.** and **C.**)
Rocky places or ravines ; where it is frequent. July 15–30.

1195. **O.** asperifolia, Michx. MOUNTAIN RICE. (**H.** and **C.**)
Woods along ravines or hills ; frequent. May 10–20.
Beside the above habitat it also occurs in the rich woods near Freeville and the Round-Marshes.

1196. **O.** Canadensis, Torr. (**C.**)
Banks of ravines ; rare. June 1–15.
Taughannock ; south side, east of the Cataract House, 1882, (*F. C. Curtice.*) Also between the Cataract House and the Falls ; and on "Eagle Cliff," above the Falls, 1884.

430. MILIUM, Linn.

1197. **M.** effusum, L. (**H.** and **C.**)
Swamps and wet places in ravines ; infrequent. June 1–20.

Fall Cr. ravine and above Forest Home. Dryden-Lansing Swamp. Malloryville. Round Marshes. Michigan Hollow and elsewhere.

431. MUHLENBERGIA, Shreb.

1198. **M. sobolifera**, Trin. (H. and C.)
Rocky places; not common. Aug.
Fall Cr. Chiefly on the lake-shore declivities, where it grows in beds. The leaves, when mature, stand at right angles from the stem.

1199. **M. glomerata**, Trin. (C.)
Sphagnum swamps, and in ravines; frequent. Aug.–Sept.
Buttermilk ravine. Larch Meadow and elsewhere. Also in dry woods on the Pinnacles near White Church.

1200. **M. Mexicana**, Trin. (H. and C.)
Shaded places; common. Aug.

1201. **M. Mexicana**, Trin. var. filiformis, Vasey, is in Fall Cr. and other ravines.

1202. **M. sylvatica**, Torr. and Gray. (H. and C.)
Moist soil; frequent. Aug.

1203. **M. Willdenovii**, Trin. (H. and C.)
Rocky places: not common. Aug.
Six Mile Creek, at the "Narrows." Enfield ravine. Thacher's Pinnacle and elsewhere.

1204. **M. diffusa**, Shreb. NIMBLE WILL. (H. and C.)
In old orchards and fields; frequent. Sept.–Oct.
University Grove, and elsewhere on East Hill. Frequent on the lakeshore points; "Sheldrake," (*Dr. Gray*, 1831.)

432. BRACHYELYTRUM, Beauv.

1205. **B. aristatum**, Beauv. (H. and C.)
Ravines and low woods; frequent. July.
The form "*var. Engelmanni*," was found in Fall Cr. by Mr. Hine.

433. PHLEUM, Linn.

1206. P. PRATENSE, L. TIMOTHY. (H. and C.)
Fields; common. June.
Since 1872 we have noticed autumnal specimens of this species with spikes bearing numerous viviparous flowering glumes ("lower palets.") It seems to be a case perfectly parallel to that of *Poa alpina*, on which Mohl based his important theory of the structure of the spikelet. The flowering glume develops into a complete leaf with sheath, ligula, and a bright green blade 1–1½ cm. long. The plants are usually growing in moist soil, and only the second growth of the season becomes viviparous. The same monstrosity appears rarely in the autumnal spikes of *Dactylis*. Viviparous *Phleums* have been observed in the Hudson R. valley by Gerard and in New Eng.

434. SPOROBOLUS, R. Br.

1207. **S. vaginæflorus**, Torr. (*Vilfa vaginæflora*, Man., p. 609. (C.)
Sandy or gravelly soil. Sept.
Near Fall Cr. mills and near Triphammer Falls. Abundant on the lakeshore banks, near Ludlowville, Aurora, Sheldrake, etc.

435. AGROSTIS, Linn.

1208. **A. perennans**, Tuck. (**H.**)
Woods and ravines ; frequent. July-Sept.

1209. **A. scabra**, Willd. (**H. and C.**)
Old logs in marshes and exsiccated places ; infrequent. July-Aug.
Eagle Hill. Michigan Hollow and other marshes.

1210. **A. vulgaris**, With. RED TOP. (**C.**)
Fields ; common. The typical form—slender and with few branches to the panicle, occurs sparingly in pastures east of Mc-Lean, in Danby, etc. The cultivated Red Top is usually much more robust.

1211. **A. vulgaris**, With., var. **alba**, Vasey, (*A. alba*,) L. (**H. and C.**)
Near the steamboat landing, Ithaca ; Lockwood's Flats and elsewhere.

436. CINNA, Linn.

1212. **C. arundinacea**, L. (**H. and C.**)
Swampy woods ; frequent. Aug.
Fall Cr. near the lake. Swamps of Freeville and elsewhere.

1213. **C. pendula**, Trin., (*C. arundinacea*, L. *var. pendula*, Gr.)
Shaded wet ravines and deep swamps ; scarce. July, Aug.
Fall Cr., in the Triphammer ravine. Near Beaver Creek. Near Locke Pond. A peculiarly delicate form with smaller flowers than the typical one occurs in Michigan Hollow Swamp.

437. DEYEUXIA, Clair.

1214. **D. Canadensis**, Beauv. (*Calamagrostis Canadensis, Beauv.*)
BLUE-JOINT. (**H. and C.**)
Open marshes ; common. June 20-July 10.
Inlet Marshes. In Dryden and Danby. Abundant on the Cayuga Marshes. Occasional in ravines. The awn arises from much below the middle of the palet in specimens from Farley's Point.

[**D. confinis**, Kth., (*C. confinis*, Nutt.) "Penn Yan," *Sartwell*, *H. and C.*]

1215. **D. Porteri**, Vasey, (*C. Porteri*, Gr.)
Local and very sparingly on Thacher's Pinnacle near W. Danby, in woods of Rock Oak, Hickory, with Vacciniums, Azaleas, etc., 1881. These specimens are less robust than the type ; but after a careful examination of plants, from our locality and from Pennsylvania, and an interesting correspondence with Prof'r Porter, the discoverer of the species, and with Dr. Vasey, to both of whom specimens were submitted, there remains but little doubt that *D. Porteri* is established as a N. Y. plant. It has hitherto been found only in a limited section of Pennsylvania, but Prof. Porter, who has collected more of it than any one else, agrees that our specimens should be referred to his species.

438. DESCHAMPSIA, Beauv.

1216. **D. flexuosa**, Vasey. (*Aira flexuosa*, L.) HAIR GRASS.
(**H. and C.**)
Dry or rocky slopes ; scarce. June 20-July 20.

The Pinnacles above White Church valley. Cayuga L. on cliffs north of Kings Ferry ; on Utts Point ; north of Aurora.

1217. **D. cæspitosa**, Beauv. (*Aira cæspitosa*, L.)

Near limestone or marl ; rare. July 1–20.

Marl Creek meadows and half mile west. Farley's and Utts Points, Cayuga L.

HOLCUS LANATUS, L., grew for several years on the south part of the Campus. (" Gorham," *Sartwell, H. and C.*)

439. TRISETUM, Torr.

1218. **T. palustre**, L. "Ithaca," (*Sartwell, Herb. Ham. Coll.*) The specimen is a genuine *T. palustre*, but the species is not now known here.

440. AVENA, Linn.

1219. **A. striata**, Michx.   (C. by Dr. Wright.)

Ravines and swamps, where it is not uncommon. May 20– June 20.

A. SATIVA, L. COMMON OAT. Often seen on Cayuga L. shore.

441. DANTHONIA, DC.

1220. **D. compressa**, C. F. Austin.

Open woods or shores ; not uncommon. July.

Dart Woods. Turkey Hill. Near Danby. Shore of Dryden L. Shore of Locke Pond. Distinguished by its usually long, flat leaves sometimes overtopping the culm, and by the awned lateral lobes of the flowering glume.

1221. **D. spicata**, Beauv.   (H. and C.)

Dry banks and hills ; common. June.

ELEUSINE INDICA, Gaertn., was found, 1874, on Aurora St., Ithaca ; not permanent. " Waterloo, N. Y." (*Sartwell, Paine's Cat.*, p. 168.)

442. DIPLACHNE, Beauv.

1222. **D. fascicularis**, Benth. (*Leptochloa fascicularis*, Gray.)

Brackish or saline soil ; rare. Aug.–Sept.

" Cayuga Marshes," (*Dr. C. H. Peck.*) Montezuma, near the salt well, also in pasture, near Salt Cr.

443. PHRAGMITES, Trin.

1223. **P. communis**, Trin. REED.   (H. and C.)

The larger marshes ; Oct. Sparingly on Summit Marsh ; but abundant throughout the Canoga, Cayuga and Montezuma Marshes. The handsome plumes of fruit become conspicuous about the middle of Oct.

444. EATONIA, Raf.

[E. obtusata, Gray ; " Seneca Lake, *Gray*,"—see Paine's Cat. p. 168 ; and " Yates Co.," *Sartwell Herb.*]

1224. **E. Dudleyi**, Vasey, nov. sp.

" Culms 2 to 2½ ft. high, very slender ; cauline leaves only 1 to 2 inches long, abruptly acute, spreading ; the radical ones 3 to 6 inches long ; panicle slender, nearly linear, 3 to 6 inches long, the branches in twos or threes below, and mostly appressed ; the upper empty glume obovate, obtuse, broadly scarious on the margins, and reach-

ing to the middle of the second floret, smoothish ; the lower glume
broader than in *E. Penn.* and nearly as long as the first floret ;
flowering glumes linear-oblong, obtuse or abruptly acute, the second one hispidulate ; palet scarious, bifid at the apex."

"Easily distinguished from *E. Pennsylvanica* by the slender culms
and panicle, the very short cauline, leaves, the longer and wider
lower glume, the more obtuse upper one, and the shorter, obtuser
flowering glumes. Grows in *dry open woods.*"

"Michigan, New York, Long Island, Penn., D. C., Va. and N.
Carolina."

The above species was collected in Cascadilla Woods in 1876 and
referred to *E. Pennsylvanica.* Again collected there in June, 1881,
its peculiarities were first observed seriously. In 1882 Mr. Curtice
obtained it from South Hill, and it occurs in both dry and moist
soil on most of our wooded ravine-slopes ; occasionally in moist
soil in more open places. Beside the range given above by Dr.
Vasey, the writer has specimens from near Pittsburg, Pa., and it is
probably in all the larger herbaria under other names.

On consulting Dr. Torrey's Herb. at Columbia Coll. in 1883, in
order to determine the position of this form, it was found,—together
with genuine *E. Pennsylvanica*, referred to the latter species ; and
it is now evident that Torrey's description of *E. Pennsylvanica* in
the Flora of N. Y., II, p. 469, was almost wholly founded on this,
and not on the true form.

1225. **E. Pennsylvanica,** Gray. (H. and **C.**)
   Marshes, shores and wet cliffs ; frequent. June 20–July.

445. ERAGROSTIS, Beauv.

1226. **E. reptans,** Nees.
   Wet sandy shores ; not common. July–Sept.
   Eddy Pond, Six Mile Cr. Fall Cr. Cayuga Lake near Ludlow-
ville and elsewhere. "Cayuga L., Gray," (*Sartwell, H. and C.*)

1227. **E. MAJOR,** Host. (*E. poæoides,* var. *megastachya,* Man., p. 631.)
   (H. and **C.**)
   Roadsides ; occasional. Aug.–Sept.
   Near Armory. Streets of Ithaca. Near the Nook. Near Coy
Glen.

1228. **E. capillaris,** Linn. (**C.**)
   Dry banks ; infrequent. Aug.–Sept.
   Near Fall Cr. Mills. North of the Nook. Mouth of Salmon
Creek ravine. Utts Point. A very small reduced form was found
in the dry soil of the Circus Common.

446. DACTYLIS, Linn.

1229. D. GLOMERATA, L. (H. and **C.**)
   Fields and borders of woods ; common. June.
   Specimens with proliferous flowering glumes found on Buffalo St.,
Ithaca, Oct., 1884.

CYNOSURUS CRISTATUS, L., was collected on Cascadilla lawn, 1878–
1885.

1230. **P. annua**, L.   (**H.** and **C.**)
Roadsides and yards; frequent. May-Oct.

1231. **P. compressa**, L.   (**H.** and **C.**)
Dry banks and woods; common. June-Aug.
Torrey's *var. sylvestris* is frequent in woods.

1232. **P. serotina**, Ehrh.   (**H.** and **C.**)
Wet meadows, usually; frequent. June-July.
It occasionally appears in dry woods, growing singly. High
banks near Ithaca Falls. Case. Woods. High hills in Danby, etc.

1233. **P. pratensis**, L. Ky. Blue-Grass.   (**H.** and **C.**)
Fields; common. June.

1234. **P. trivialis**, L.
Marshes and fields, and the deep swamps. June.
From the large number of swamp stations, it is probably indige-
nous in this region.

1235. **P. sylvestris**, Gray.
Slopes of ravines; rare. June.
Below Green-Tree Falls—the genuine form.

1236. **P. sylvestris**, Gray, var. palustris.
Sphagnum bogs and cold swamps. Round Marshes. Michigan
Hollow. Wyckoff swamp. Culm weak and slender, branches of
the panicle, in twos and threes deflexed, otherwise like the type.
A much larger rather robust form grows in Round Marshes, and in
Dryden-Lansing Swamp.

1237. **P. debilis**, Torr.   (**H.**)
Dry woods; frequent. May 20-June.

1238. alsodes, Gray.   (**H.** and **C.**)
Ravines—Fall Cr., Six Mile Cr., and especially in the deep
swamps, like Beaver Cr. and Freeville swamps.

448. Glyceria, R. Br.

1239. **G. Canadensis**, Trin. Rattlesnake Grass.
Marshes; rare. Summit Marsh. "Junius," (*Sartwell, H. and C.*)
July.

1240. **G. elongata**, Trin.; doubtful and overripe specimens from north
of Freeville; and west of Locke Pond, (*F. C. Curtice*,) probably
belong here.

1241. **G. nervata**, Trin.   (**H.** and **C.**)
Marshy ground; common. May-June.
This species is variable, the purple-flowered specimens with large
spikelets, appear quite different from those with the smaller green
spikelets.

1242. **G. pallida**, Trin.   (**H.** and **C.**)
Marshes in water; not common. June.
Near the Corner-of-the-lake. Cayuta L. Danby, in Jennings
Pond.

1243. **G. arundinacea**, Kth. (*G. aquatica*, Man., p. 627.)   (**H.** and **C.**)
Wet meadows; common. June.

1244. **G. fluitans**, R. Br. (H. and C.)
Ditches and pools ; not common. June.
About Ithaca. Summit Marsh. Dryden-Lansing Sw., and elsewhere.

1245. **G. acutiflora**, Torr.
Standing pools and marshes. June.
"Ithaca," (*Sartwell, H. and C.*) Marsh near the lake. South Hill Marsh, etc. More abundant than the preceding species.

449. FESTUCA, Linn. FESCUE GRASS.

1246. **F. tenella**, Willd. SLENDER FESCUE. (H. and C.)
Banks of ravines and dry woods ; frequent. June.
Casc. and Fall Cr. Woods and elsewhere. Awns often ⅓ longer than the flowering glume instead of "shorter or equalling" it.

1247. **F. ovina**, L.
Lawns (introduced) in the University Grove, on State and Seneca Streets and elsewhere. In woods (possibly native) north of King's Ferry, on Utts Point and banks of the lake north of Union Springs and of Aurora. May–June.

1248. F. DURIUSCULA, Linn. (C.)
Lawns ; in the University Grove ; Casc. Place. Utts Point.

1249. F. ELATIOR, L. (H. and C.)
Fields ; common. June.

1250. **F. nutans**, Willd. (H. and C.)
Woods ; frequent. June.
In Casc. Woods and generally distributed throughout our limits.

450. BROMUS, Linn.

1251. B. SECALINUS, L. CHESS. (H. and C.)
Fields and roadsides ; frequent. June.

1252. B. RACEMOSUS, L.
Scarce ; near McLean, 1879 ; near Lehigh Valley depot ; near Lick Brook. Aurora.

1253. **B. Kalmii**, Gray.
Ravines and open meadows ; scarce. June.
Six Mile Cr., above the Sulphur Spring, and near Green-Tree Falls. Larch Meadow. Near Beaver Cr.

1254. **B. ciliatus**, L. (C.)
Ravines and river banks ; abundant. June–July.

1255. B. ciliatus, L. var. purgans, Gr., is in Turkey Hill Woods ; near Ludlowville and elsewhere.

1256. **B. ciliatus**, L. var. ——, approaching some of the Rocky Mt. forms (according to Dr. Vasey,) occurs in our sphagnum bogs or wet meadows. The plants are low, light-green ; the panicle peculiarly chaffy in appearance, light-colored, and the flowering glumes smooth on the back but strongly ciliate. It is abundant, in Round Marshes, along Locke Pond and elsewhere.

1257. B. STERILIS, L. (H. and C.)
Rare ; June. Taughannock near the landing, 1884, (*F. V. Coville.*) !

1258. L. PERENNE, L. DARNEL.

Lawns and fields ; local. June-Oct.

Near Casc. Place since 1879. Field south of "Edgewood," rather abundant. E. of Dr. Law's. "Geneva," (*Sartwell, H. and C.*)

HORDEUM MURINUM, L., appeared one year, 1882, on the Fiske McGraw grounds, (*F. L. K.*)!

HORDEUM JUBATUM, L., grew on the south part of the Campus, 1878–1881.

452. AGROPYRUM, Beauv.

1259. A. repens, Beauv., (*Triticum repens, L.*) COUCH GRASS.
(H. and C.)

Roadsides, and dry fields ; frequent. July.

On the Campus, near the "Diamond," is a bed of low, glaucous plants, almost wholly awnless.

1260. A. caninum, R. and S., (*T. caninum, L.*)   (H. and C.)

Dry woods and banks of ravines ; frequent. July.

A variable species, often approaching *T. violaceum* in the character of the awns. A very slender straight-awned form is found on the high wooded hills of Danby and Caroline.

1261. A. caninum, R. and S., var. ———, occurs near Lowery's Pond in the marsh, and approaches *T. violaceum*, in the somewhat purplish flowers, the erect spike and short and straight or even spreading awns.

453. ELYMUS, Linn. RAY-GRASS.

1262. E. Virginicus, L.   (H. and C.)

Dry banks of ravines, and river-banks ; frequent. July-Aug.

Flowers are often hairy, instead of "smooth."

1263. E. Canadensis, L.   (H. and C.)

River-banks and shores ; frequent. July.

1264. E. Canadensis, L., var. glaucifolius, Gr.

Especially abundant at the foot of the cliffs all along the lake shore.

1265. E. striatus, Willd.   (H. and C.)

Fall Cr. Casc. Cr. Enfield ravine. Cayuga L. shore.

1266. E. striatus, Willd., var. villosus, Gr.

Negundo Woods. Near Salmon Cr. Near Shurgur's Glen and elsewhere near the lake ; not rare.

454. ASPRELLA, Willd.

1267. A. Hystrix, Willd., (*Gymnostichum Hystrix*, Schr.)
(H. and C.)

Hedges and borders of woods ; frequent. July.

SECALE CEREALE. Willd. RYE. Frequently springs up by roadsides.

GYMNOSPERMÆ.

CONIFERÆ.

455. THUYA, Tourn.

1268. **T. occidentalis**, L. ARBOR-VITÆ. WHITE CEDAR.
(**H.** and **C.**)

In "cedar swamps," or marshes; scarce. May 10–20.

A half mile northwest of Black Lake, in a swamp north of Lay's Iron Spring,—a large number. Scattered trees occur about Lowery's Pond.

456. JUNIPERUS, Linn.

1269. **J. communis**, L. JUNIPER. (**C.** by Dr. Wright.)

Rare; May 1–15. Three stations are known: In a pasture west of Eagle Hill; South Hill, north of S. S. 420; W. Danby, near the western base of Thacher's Pinnacle, (*F. V. Coville.*) As the plants in all cases are single and staminate, possibly the species is not truly indigenous here.

1270. *****J. Virginiana**, L. RED CEDAR. (**H.** and **C.**)

Rocky banks or hillsides; not uncommon. Apr. 15–30.

Fall Cr. and other ravines, and abundant along the banks of Cayuga L. A fine tract of them on the hillsides south of Larch Meadow.

457. TAXUS, Tourn.

1271. **T. Canadensis**, Willd., (*T. baccata,* L., var. *Canadensis,* Man., p. 474.) GROUND HEMLOCK. YEW. (**H.** and **C.**)

Along ravines and in cold swamps; frequent. Apr. 20–May 10.

Among the swamp habitats are Ellis Hollow Swamp, Michigan Hollow and Bear Swamp.

458. Pinus, Tourn.

1272. **P. strobus**, L. WHITE PINE. (**H.** and **C.**)

Woods and hills; frequent. June.

Once the principal forest tree over large areas in this vicinity. Occasionally single first-growth pines are still seen, rising far above the surrounding forest. In Rhodes Woods there are many such; but the only undisturbed tract is "Signer's Woods," about forty acres in extent, largely of pines, and lying in the bottom of the valley north of Summit Marsh. There are many magnificent trees $130°$–$150°$ in height.

1273. **P. resinosa**, Ait. RED PINE. NORWAY PINE. (**H.** and **C.**)

Dry banks and declivities; not common. June.

North bank of Six Mile Cr., below Well Falls; on the promontory east of the Sulphur Spring. Mouth of Coy Glen. Frequent on north bank of Buttermilk ravine. High bank north of Lucifer Falls. Abundant on the high ridge east of W. Danby. On Ball Hill. The largest groups are on the declivities between White Church and Brookton. East shore of Cayuga Lake, from McKinney's to Ludlowville. Wanting on the west shore except at Taughannock and Trumansburg ravines.

1274. **P. rigida**, Mill. Pitch Pine.   (**H**. and **C**.)
  Hills or dry soil; frequent. June.
  Campus, Cascadilla Woods and elsewhere.

459. Picea, Link.

1275. **P. nigra**, Link. (*Abies nigra*, Poir.) Black Spruce. (**C**.)
  Scarce; May. Enfield "Spruce Swamp." Woodwardia Swamp,
  where the trees are dwarfed.

460. Tsuga, Endl.

1276. *****T. Canadensis**, Carriere. (*Abies Canadensis*, Michx., Hemlock.   (**H**. and **C**.)
  Ravines and swampy woods; common. June.
  The central portion of Round Marsh Woods is largely made up
. of first-growth hemlocks, where some of the trees attain splendid
  proportions.

461. Abies, Tourn.

1277. **A. balsamea**, Miller. Fir. Balsam Fir.   (**H**. and **C**.)
  Scarce; May. Fir-Tree Swamp northeast of Freeville. Also a
  few trees on West side of S. C. railroad. Near the mouth of Mud
  Creek, formerly. Fir-Tree Swamp southeast of Danby village.

462. Larix, Tourn.

1278. **L. Americana**, Michx. Larch. Tamarack.   (**H**. and **C**.)
  Marshes and swamps; infrequent. May 1–15.
  Larch Meadow formerly; still among the cedars south. Freeville,
  (*Dr. Jordan.*)! Mud Cr. Woodwardia Swamp. Fir-Tree Swamp
  east of Danby. Michigan Hollow,—large trees. W. Junius, Tama-
  rack Swamp and near the ponds.

Additions and Corrections.

p. vii.   For "*Aster sagithfolius*," read "*Aster sagittifolius.*"

p. xi., line 16.   For "plant" read "plants."

p. xiv., line 20.   For "*Lathyrus palustris*" read "*Lathyrus paluster.*"

p. xv., line 7 from bottom.   For "eastern" read "south-eastern."

p. 2.   Erase all under "13ª."

p. 3.   For "*R. alismæfolias*," read "*R. alismæfolius.*"

p. 12, no. 94.   For "*P. graveolens*" read "*P. graveolens.*"

p. 12, no. 96,   For "*L. Majoi*" read "*L. major.*"

p. 14, no. 107.   For "sepels" read "sepals," and for "placenta" read "placentæ."

p. 19.   Before genus "72, AILANTHUS" write Order
"SIMARUBACEÆ."

p. 24, no. 197.   For "*Robinia Psevdacacia*" read "*R. Pseudacacia.*"

p. 26, no. 223 and no. 224.   For "*Lathyrus palustris*" read "*Lathyrus paluster.*"

p. 27, no. 229.   For "*G. triacanthus*" and "*G. triacanthos.*"—

p. 42, no. 381.   For "*N. multiflora*" read "*N. sylvatica*," Marsh.

p. 46, line 3.   For "*Dipcacus*" read "*Dipsacus.*"

p. 58, no. 572.   After "S. S. 420" add : "also 1 mile southwest of W. Junius."

p. 67, no. 657.   For "*S. nodosa*, L." read "*S. nodosa*, L., var. *Marilandica*, Gray."

p. 68, no. 672.   For "*G. Pedicularia*" read "*G. pedicularia.*"

p. 72, no. 710.   For "*M. Clinopodia*" read "*M. clinopodia.*"

p. 76, no. 743 and aud 744.   For "*B. rubrum*" and "*B. Capitatum*" read "*Chenopodium rubrum*, L." and "*C. capitatum*, Wats."

p. 99, no. 928.   For "*M. Virginica*" read "*M. Virginiana.*"

p. 117, no. 1134.   Strike out "var. *elongata*, Torr."

p. 124, no. 1206.   For "in Hudson R. valley, by Gerard," read "on Staten Id, by Britton and Hollick.

Index to Orders and Genera.